**Chemical Regulation
in the Middle East**

Chemical Regulation in the Middle East

Michael S. Wenk, M.S.
Bergeson & Campbell, P.C.
The Acta Group
Washington, D.C., USA

This edition first published 2018
© 2018 John Wiley & Sons Ltd

The right of Michael S. Wenk, Bergeson & Campbell, P.C., and The Acta Group to be identified as authors of this work has been asserted in accordance with law.

Registered Offices
John Wiley & Sons, Inc., 111 River Street, Hoboken, NJ 07030, USA
John Wiley & Sons Ltd, The Atrium, Southern Gate, Chichester, West Sussex, PO19 8SQ, UK

Editorial Office
The Atrium, Southern Gate, Chichester, West Sussex, PO19 8SQ, UK

For details of our global editorial offices, customer services, and more information about Wiley products visit us at www.wiley.com.

Wiley also publishes its books in a variety of electronic formats and by print-on-demand. Some content that appears in standard print versions of this book may not be available in other formats.

Library of Congress Cataloging-in-Publication Data applied for
Hardback: 9781119223641

Cover design: Wiley
Cover image: © motorolka/Shutterstock;
           © TonelloPhotography/Shutterstock

Set in 10/12pt WarnockPro by SPi Global, Chennai, India
Printed in Singapore by C.O.S. Printers Pte Ltd

10 9 8 7 6 5 4 3 2 1

To Cynde

Contents

Introduction

Historically, chemical regulation in the Middle East has been a national patchwork of regulations, with the substances and applications potentially experiencing varying types of requirements. Indeed, "the level of sophistication of chemical control regulations varies greatly as a result of different levels of economic and industrial development."[1] At one end of the spectrum lies Israel, which has a fairly comprehensive system of such regulations in place, often mirroring closely those of the United States and/or the European Union. At the other end can be found countries such as Pakistan, where such laws are perhaps more the exception than the rule. Indeed, "the level of sophistication varies greatly as the result of different levels of economic and industrial development."[2]

While many of the countries to be discussed have exhibited an intent, if not a formal action plan, to develop such schemes, they are normally less advanced in this regard than entities such as the United States, the European Union, and various Asia-Pacific-region countries such as Japan, Australia, and South Korea. Generally, there are no national chemical inventory programs in the Gulf States region; instead, such countries tend to focus more on a system of permits and licenses to manage chemicals. There are, however, often a wide variety of Ministry, legislative, and other decrees which relate to pesticides, Occupational Safety and Health, Safety Data Sheets, and product labels, again to widely varying degrees.

Before moving forward, readers should be aware of two key issues with respect to how regulations are referenced in many of the countries of the Middle East. First, with respect to the date format employed, "Gulf countries typically use the Islamic calendar for religious purposes, but the Gregorian for secular purposes."[3] Indeed, "[t]he Islamic calendar has 12 months, each with 29 or 30 days. The Islamic calendar year is 11 days shorter than the Gregorian calendar year."[4] It is not uncommon to see both the Islamic calendar and the Gregorian calendar date presented when examining a particular regulation from the region. For example, the Egyptian "Prime Minister's Decree No. 338 of the Year 1995" is internally cited as having been "Issued in the Cabinet on 18 Ramadan Hejira Year 1415 (Corresponding to February 18, 1995)."[5] Second, regulations often present the date of promulgation using only the year, and do so in two different ways. The more common format is to display the regulation name, followed by the year of enactment in parentheses, such as Legislative Decree No. 11 (1989). An alternative way of displaying a regulation title in the region is to note the date of enactment after the title and following a / symbol, such as Royal Decree 84/2011. These formats tend to vary from country to country, but are generally consistent within a given country.

Acknowledgments

This work would not have been possible without the extremely generous support of Lynn L. Bergeson, President, The Acta Group, and Lisa M. Campbell, Partner, Bergeson & Campbell, P.C.

I would also like to thank Elsie Merlin, Project Editor, Physical Sciences and Engineering, and Emma Strickland, Assistant Editor, as well as all of the technical reviewers at John Wiley and Sons, Ltd. for their dedication, patience, and guidance.

Thank you to my parents, whose love and guidance are with me in whatever I pursue.

Finally, I wish to thank my best friend, my *ana mandara* – "beautiful refuge" – my wife Cynde for her unwavering support, love, and encouragement.

1

The Cooperation Council for the Arab States of the Gulf

National Overview

The Cooperation Council for the Arab States of the Gulf (alternatively the "GCC" or the "Council") is a cooperative framework, launched on May 25, 1981 (21st Rajab 1401 AH in the Islamic calendar), "to effect coordination, integration and inter-connection among the Member States."[6] The GCC, headquartered in Riyadh, Saudi Arabia, is comprised of six Member States: the United Arab Emirates (which itself is composed of Sharjah, Ajman, Dubai, Abu Dhabi, Fujairah, Ras al-Khaimah, and Umm al-Quwain), Kuwait, Saudi Arabia, Bahrain, Qatar, and Oman. Each of these entities is an autocratic monarchy or sheikhdom and, consequently, has little to no citizen participation in their political processes. "All GCC members are also members of the Arab League and Qatar, Saudi Arabia, Kuwait and the United Arab Emirates are prominent members of OPEC."[7]

Governmental Structure

The GCC is vested with the authority to issue mandatory laws and technical regulations, as well as voluntary standards. However, it should be noted that sovereignty resides with each Member State, which may or may not choose to adopt these types of regulatory devices.[8] A loose parallel may be seen with regard to the European Union (EU), where regulations become immediately enforceable as law in all Member States simultaneously, while directives must be passed into national law by each individual Member State. Further, GCC Member States are also entitled to introduce more stringent requirements at the national level, on condition that they adhere to the GCC framework.[9] The GCC is governed by a series of agencies, as follows:

1. The Supreme Council, to which shall be attached the Commission for Settlement of Disputes;
2. The Ministerial Council; and
3. The Secretariat General.

The GCC has established a nascent defense force (1984), developed an intelligence sharing agreement (2004), and created a common market similar to that of the EU in

Chemical Regulation in the Middle East, First Edition. Michael S. Wenk.
© 2018 John Wiley & Sons Ltd. Published 2018 by John Wiley & Sons Ltd.

2008. However, as with the EU, a single currency adopted by all Member States was not established.[10] Indeed,

> The decision [to form such an alliance] was not a product of the moment but an institutional embodiment of a historical, social and cultural reality. Deep religious and cultural ties link the six states, and strong kin relations prevail among their citizens… Therefore, while, on [the] one hand, the GCC is a continuation, evolution and institutionalization of old prevailing realities, it is, on the other [hand], a practical answer to the challenges of security and economic development in the area. It is also a fulfilment of the aspirations of its citizens towards some sort of Arab regional unity.[11]

According to the "Charter of the Gulf Cooperation Council" (Charter), enacted on May 25, 1981, the goals of the GCC are:[12]

1. To effect coordination, integration, and inter-connection between Member States in all fields in order to achieve unity between them.
2. To deepen and strengthen relations, links, and areas of cooperation now prevailing between their peoples in various fields.
3. To formulate similar regulations in various fields, including the following:
   A. Economic and financial affairs;
   B. Commerce, customs, and communications;
   C. Education and culture;
   D. Social and health affairs;
   E. Information and tourism; and
   F. Legislative and administrative affairs.

The 1981 "Unified Economic Agreement Between the Countries of the Gulf Cooperation Council" (Agreement) requires Member Countries to "coordinate activities in the areas of trade, commercial [development], customs, financial, import, export, transportation, communication policies, professional and technical training, as well as in projects facilitating overall economic development."[13]

Key Chemical Regulatory Agencies

The Standardization Organization of the Cooperation Council for the Arab States of the Gulf, also known as the GCC Standardization Organization (GSO, or Organization), is also headquartered in Riyadh. Generally speaking, the function of the Organization is to develop GCC-wide standards for a variety of aspects, including but not limited to: water, fire protection, Halal foods, cigars and cigarettes, and food safety. Where no GSO standard already exists, the Organization may elect to review existing global standards and to adopt them, or permutations thereof, as it sees fit.

One example of such a standard implemented by the Organization is the "GSO Technical Regulation on Toys: Second Edition – BD-131704-01" from November 27, 2007. This technical regulation was adopted by the GSO Board of Directors at its 17th meeting (May 31, 2013) and entered into force on January 1, 2014.[14] Specific to chemical regulation, the GSO Technical Regulation on Toys has required the amendment of the essential requirements, especially with regard to the prohibition of the use of certain chemicals that cause cancer or genetic defect or allergies or concerning the use of fragrances, as well as the maximum limits permitted for certain substances especially in toys intended for use by children under thirty-six months or toys that children can put in their mouths.[15]

Key Chemical Substance Regulations

In 1997, the GCC published the "General Regulations of Environment in the GCC States" (Regulation), recognizing that "The great developmental growth, in the last few years in GCC States, has created some negative results, which are considered a clear threat to the environment in the area and calls for positive attention and consideration."[16] This regulation, while specific in title to the environment, has implications with respect to the management of chemical substances in the GCC.

While some of the language of Article 1 of the Regulation may be viewed as fairly grandiose (e.g. "Every person has a basic right to live a convenient life in an environment compatible with human dignity"), Section 4 lays out the specific responsibility for the management of the environment, stating

> The responsibility for management of the environment, its natural and wildlife resources, particularly the ability of the natural resources to satisfy the development needs of the present and future generations, lies on the shoulders of the official authorities, public and private establishments as well as responsible and ordinary persons.[17]

Specifically, Section 7 directs

> [e]nvironmental considerations should be taken into consideration and given foremost priorities. These priorities should be merged with all stages and levels of planning so as to make environmental planning an integral part of the comprehensive development planning in all industrial, agricultural, constructional and other field [sic]…[18]

As a result, consideration of the natural environment as early as possible in the planning stages of a wide variety of areas is paramount. Section 10 focuses new facilities and projects on the use of Best Available Technology (BAT) with respect to controlling pollution and preventing environmental degradation, as well as directing existing projects to "use the technologies that guarantee compliance with environmental feasibility standards… or the technologies that prevent the occurrence of any substantial negative effect on the environment."[19] As an aside, Article 6 of the Regulation speaks more specifically to the use of BAT.

Article 3 of the Regulation lays out the "Duties of the Public Bodies" with respect to environmental management, specifically:[20]

1. Working to prevent negative environmental effects which may result from their own projects or projects under their supervision or licensed by them;

2. Taking all suitable measures to guarantee application of the rules stipulated in this regulation on their own projects under their supervision or licensed by them, including compliance with the environmental protection regulations and standards in force, in addition to issuing the necessary rules, regulations, and additional guidelines in this regard in consultation with the concerned authority;

3. Monitoring the application of the environmental regulations and standards and compliance with them on their own projects or projects supervised or licensed by them and preparing periodic reports about their effectiveness and the extent of compliance to them; and

4. Any public body responsible for issuing standards, specifications, or rules related to activities affecting the environment should give the concerned authority ample opportunity to give its opinion before granting such licenses or permits or issuing such regulations, standards, or specifications.

Article 4 defines the "Duties of Persons" with respect to the same:[21]

1. Any person responsible for the design or implementation of any project should make sure that the design and operation of such a project complies with the regulations and standards of environment protection;

2. Any person intending to do any work, or neglecting to do any work, that may lead to the occurrence of negative effects on the environment should seek, through studying the environment assessment or any other suitable means, to identify those probable effects and take all suitable measures to prevent such negative effects or reduce their probability to the lowest possible level;

3. In case of the occurrence of any of the probable negative effects, the concerned person should take the necessary steps to stop or alleviate such effects; and

4. The person responsible for any work that causes damage to the environment, or a negative effect on the environment caused by his neglect, shall not be absolved from the responsibility for the damage caused to the environment as a result of his actions or negligence even after complying with the provisions of Paragraph (2) of this Article.

Related to the foregoing, Article 7 recommends that those individuals who have a supervisory position and/or role in projects which may have "intense negative effects" on the environment should designate an employee whose role will be to ensure compliance with the Regulation, as well as "any other regulations issued by virtue of it," thus further cementing the role of environmental protection efforts in the types of projects and sectors addressed previously.[22] Subsequent Articles discuss the management of land, and specifically of coastal areas (Article 9, "Land Uses and Coastal Zones"), management of species, both animal and plant, as well as their attendant habitats (Article 10, "Conservation of the Living Species"), and the maintenance, management, and further development of renewable and non-renewable resources (Article 11, "Rationalization of the Use of Natural Resources") and environmental education at both primary and secondary school levels, and also in the technical training environment (Article 12, "Environmental Education and Environ-mental [sic] Awareness").[23]

The final four Articles of the Regulation relate in various forms to the additional requirements the authority may impose on the person conducting the project (Articles 17 and 19), the penalties which may be levied by those entities failing to comply with the requirements set out (Article 18), and the right of the authority to inspect projects at its discretion (Article 20).

Article 17 ("Application of Environmental Protection Standards") grants the responsible authority the ability to require a specific set of data if "in the opinion of the concerned authority, [it] may produce pollutants or lead to deterioration of the environment."[24] Such information may include the nature of the activities and the materials used, the wastes from the activities, and/or the methods or equipment used to reduce or stop the resultant pollution or environmental degradation.[25]

Article 18 ("Infringement of the Environmental Laws") directs the concerned authority, once it "becomes certain" – interestingly, not simply on the basis of suspicion – to ask the responsible person to submit a report detailing the steps that will be undertaken in order to prevent the infringement from reoccurring.[26] The responsible authority must approve the steps, and may ask the person to both "remove the solid, liquid, or gaseous wastes or any other negative effects" caused, and "to restore the situation to its original condition before the occurrence of the infringement whenever possible."[27] In situations where failure to comply with attendant obligations has been observed, the responsible authority may additionally require the suspension of work on the project which has caused the non-compliance event(s), until appropriate measures are taken.[28]

Article 19, similar in concept and scope to Article 17, grants the authority the ability to "take the measures it deems necessary to avoid, prevent, or reduce damage to the lowest possible level before it occurs."[29] This ability may manifest itself by directing a work stoppage (temporary or permanent), imposing restrictions on the work activities which are to be performed, imposing technical, operational, or other standards, and/or "any other means which the concerned authority deems fit."[30]

Finally, Article 20 ("Inspection") entitles the authority concerned to enter facilities as it sees fit, and to request and remove "samples from refuse and the materials used or stored in the project or produced by it," to verify compliance with applicable regulations and standards.[31]

While the Regulation sets forth a baseline effort, the ensuing 2002 "Common System for the Management of Hazardous Chemicals in the Gulf Cooperation Council for the Arab States of the Gulf" (Common System) laid out the minimum legislative requirements for each Member State for managing hazardous chemicals; again, for those entities which chose to incorporate them into national law.[32] The Common System contained no provisions for a GCC-wide chemicals notification or similar management scheme.[33] The GCC does, however, ban the sale and consumption of substances defined as ozone-depleting under the Montreal Protocol.[34] Further, there is no overarching regulation or law at the GCC level to manage Volatile Organic Compounds (VOCs). These compounds are regulated mainly through standards and technical rules.[35]

Article I of the Common System delineates those substances which are considered to be "hazardous chemicals" as:

> Chemicals in gaseous, liquid or solid form(s) as referred to in the category lists (attached hereto) and characterized to be effective, toxic, explosive, corrosive or other characteristics that can result in a hazard to human health and environment, whether on their own or when in contact with other materials.[36]

"Environmental hazards" are defined to be "Direct and accumulated hazards that occur in water, air and soil, can potentially pose a risk to man, flora and fauna, cause damage to living resources and ecosystems, and minimize other habitual uses of environmental sources, individually or combined."[37] Annex I of the Common System further lists the specific classifications of hazardous chemicals. Generally speaking, these classifications mirror the United Nations Recommendations on the Transport of Dangerous Goods (UN Recommendations) classifications of the same, which are contained in the UN Model Regulations prepared by the Committee of Experts on the Transport of Dangerous Goods of the United Nations Economic and Social Council (ECOSOC):[38]

- Category 1: Explosive materials
- Category 2: Compressed or liquid gases
- Category 3: Flammable liquids
- Category 4: Flammable solid materials
- Category 5: Oxidizing agents
- Category 6: Toxic materials
- Category 8: Corrosive materials
- Category 9: Other hazardous materials

Readers should note, however, that Annex I does not include the UN Recommendations Category 7: Radioactive material. In addition, as stated above, Annex I only largely follows the UN Recommendations, especially with respect to classification. One example of this may be seen when comparing the Common System definition of a "flammable solid" with the UN Recommendations definition of the same, respectively:[39]

A. Flammable solid materials:
1. Are readily combustible and can cause fire through friction.
2. Are in powdery, granular or paste form, and, if easy to burn on contact with a source of combustion, can be hazardous. Hazard can be formed by fire and combustion toxic products as well.
3. Fires caused by metal powders are difficult to extinguish, hence the particular hazard they form. Ordinary extinguishing materials such as carbon dioxide and water can increase risk.

versus

Division 4.1 Flammable solids

Solids which, under conditions encountered in transport, are readily combustible or may cause or contribute to fire through friction; self-reactive substances and polymerizing substances which are liable to undergo a strongly exothermic reaction; solid desensitized explosives which may explode if not diluted sufficiently.[40]

Article II of the Common System, "Scope of Application," sets out that the regulation shall apply to: [41]

[a]ll practices which comprise hazardous chemicals management, including:
a. [p]roduction and various (industrial, agricultural, veterinary) uses of hazardous materials for educational, research or training activities, or in any other activity associated with a certain use, which involves dealing with chemicals [,]
b. [a]ny other practices determined by the competent regulatory authority in the state, [and]
c. [e]xcept for drugs and narcotics for medical purposes, radioactive materials, explosives and weapons.

Thus, while the Common System appears to exempt the "normal" applications and uses for chemical substances, similar to the chemical legislation in various other countries outside the GCC – medical purposes, radioactive materials, explosives, and weapons – it does, interestingly, retain control over agricultural and veterinary applications. Many of these other countries exclude such compounds from their national law of chemical substances, as they are often managed by other regulations.

Article III sets out what the Common System calls the "Basic Obligation[s]" of those to whom the System is applicable:[42]

1. Unless authorized and monitored by the competent regulatory authority, no practices or procedures involving dealing with hazardous chemicals or relevant equipment shall be applied, introduced, conducted, modified, suspended, or terminated;
2. Unless authorized and monitored by the competent regulatory authority, no hazardous chemical shall be manufactured, produced, acquired, possessed, imported, exported, purchased, sold, delivered, received, lended, borrowed, modified, circulated, used, transported, stored, ceased from operation, or discharged; [and]
3. Unless authorized by the competent regulatory authority, no site for any practices or work involving a hazardous chemical or equipment involving a hazardous chemical shall be allocated; nor shall any relevant buildings, facilities, or places be set up or modified for the same.

As discussed earlier, there is presently no chemical substance registration or other notification requirement in GCC regulations. Interestingly, though, Section 5 of Article IV ("Responsibilities of the Competent Regulatory Authority") requires the competent authority to:

> Set up national databases for hazardous chemicals as regards their chemical and physical properties and hazards, secure constant and accurate statistics [undefined] of hazardous chemicals, and periodically publish the same in journals as considered vitally important for data reporting and operational studies.[43]

It seems unusual to set up national databases only to seemingly separately record the chemical and physical properties and also the hazards of such substances, data which is freely available on the Safety Data Sheet (SDS) of such materials from a variety of sources. The "constant and accurate statistics" are, as noted, not defined specifically in the Common System, so how the competent regulatory authority actually complies with Article IV is unclear.

Article V ("Licensing") does appear to lay more formal groundwork regarding the use of chemical substances, but again stops short of delineating a specific registration scheme. Section 1 requires that "Anyone… intending to implement any of the practices and/or works associated with hazardous chemicals, set forth in Article III, shall apply to the competent regulatory authority for a work and/or practice license."[44] Such a license "shall be issued by the competent regulatory authority in the country for a specific period of time and practices."[45] Article V goes on to proscribe a prohibition against assigning the license to others (Section 4), the attendant data requirements, such as to "retain a record of the trading movement… numbered and stamped by the competent regulatory authority and retained for a five-year period…" (Section 6), and a requirement to develop contingency plans for accidents, which are to be submitted to the competent authority in the country at issue (Section 9).[46] Such licenses are discussed later, in the specific country's section.

Article VI of the Common System lays out the issues relating to the importation of products, declaring in Section 1 that no import of hazardous substances may take place "until approval of [read: by] competent regulatory authority and/or bodies concerned is obtained."[47] The approval, in the form of a license, may be granted provided the applicant provides the Authority with a Material Safety Data Sheet (MSDS) for the substance, as well as the following, at least 30 days prior to the start of the import process (Section 2):[48]

1. The chemical's scientific and trade names and its chemical composition;
2. The chemical's UN number and Chemical Abstracts Service (CAS) numerical registry number;
3. The chemical's hazard degree and health and environmental effects;
4. The weight of hazardous chemical intended for importation;
5. Transportation date and time expected;
6. Importation purpose;
7. Optimal methods for chemical's storage and disposal;
8. Actions to be taken in case of hazardous material leakage;
9. The full name, correct address, and contact number of the forwarding agent, consignor, consignee, and beneficiary destination;
10. A "Certificate of origin and testing" in the material-exporting country; [and]
11. The chemical's expiration date.

Perhaps primarily because the GCC has neither formally adopted nor mandated the "Globally Harmonized System of Classification and Labeling" (GHS), the acronym "MSDS," as opposed to the likely more familiar "SDS," is generally used throughout the Common System. Interestingly, the SDS portion of Article VI requires the "weight of hazardous chemical to be imported," the "date and time of the expected transfer," the "purpose of the import," the "full name and address and contact number of the forwarding agent and consignee," the "certificate of origin and testing," and the "chemical's date of validity" – items not normally found on SDS globally.[49] Readers should be aware that, while Annex II of the Common System provides a sample SDS, the data in the Annex differ from the requirements of Article VI.[50]

Article VII of the Common System addresses the requirements of packaging and labeling of hazardous substances. Section 1-7 requires the use of good-quality, compatible packaging materials, with specific criteria listed for liquid-form chemicals. UN and/or national packaging specifications are required.[51] Section 2-7 addresses the requirements for "Handling and Hazard Labels":[52]

1. The package shall be of a size adequate for attaching all the signs and information labels required by the Material Safety Data Sheet (MSDS) and in accordance with other national regulations. [*Note:* no mention is made in this Section as to how to manage containers for which the size may not be adequate, such as for sample quantities. Other national (non-GCC) regulations have made provisions for such situations, allowing labeling to be attached (as opposed to being printed on or affixed) to the container];

2. Labels shall be affixed to packages with a material durable enough to resist ordinary conditions of transport, in order for information therein to remain clearly identifiable, legible in *Arabic and English* [emphasis the author's] and resist wear or tear;
3. Labels shall contain informative pictographs and show in internationally approved colors warning phrase and sign in accordance with standard codes [sic]; and
4. Label [sic] shall comprise the following data:
   a. Manufacturer's name and registration number in the producing country;
   b. Production and expiry date under all storage conditions of the package;
   c. Chemical and trade names, active substance, purity ratio and impurities specifications, if any; and
   d. Necessary precautions [first aid] are required for humans and non-targeted organisms to protect and treat exposures from hazards.

Article VIII of the Common System addresses the shipment of hazardous chemicals by land (Section 1-8), air (Section 2-8), and sea (Section 3-8), and the proscriptions therein generally follow international conventions, such as the International Air Transport Association (IATA) and the International Maritime Dangerous Goods (IMDG) Code.[53] Section 4-8, "Transport by Post," expressly forbids the transport of hazardous chemicals by this route, irrespective of the three foregoing sections.[54]

Article IX of the Common System, "Storage," is among the most detailed of all the Articles, establishing in substantial detail criteria for the physical design and operation of warehouses which store hazardous chemicals, and the minimum distance from publicly frequented areas that such chemicals must be segregated (depending upon the category of the same). Interestingly, Section 4 specifies that "Hazardous chemicals shall be segregated in accordance with UN classification standards and requirements set forth in Table 2…" While a reference is not provided to the applicable "UN classification standard and requirements," in either the Article or the Section, Table 2 appears to mirror the concept of the 2014 Edition of the IMDG Code segregation table, with the Segregation Terms in 7.2.2.2 ("away from," "separated from," and so forth) being replaced with specific distances.

Article X relates to "Production and Use," and details the items which must accompany a complete application for licensing, while Article XI sets out the guidelines for Occupational Exposure Limits (OEL) for each member country. Notably, the Article does not direct which OELs are to be followed (e.g. National Institute for Occupational Safety and Health (NIOSH), American Conference of Governmental Industrial Hygienists), only that:

> Each country, guided by chemicals limits and levels listed in Annex (3) table [sic], shall establish the limits and levels of occupational and environmental exposure to hazardous chemicals. Said levels and limits may not be exceeded. Safety standards issued by the relevant international organizations concerned with hazardous chemicals *can be* utilized [emphasis the author's].[55]

Article XII lays out the authority for the competent authorities to "periodically and suddenly" inspect all activities relating to hazardous chemicals, while Article XIII specifically details the "Penalties and Sanctions" requirements member countries adopting the legislation should establish (e.g. "National laws and regulations of each country shall comprise explicit provisions for penalties of imprisonment or fine or both for each violation of Article III and Article X provisions..."), and discusses the situations in which non-compliance with the license provisions is observed.[56]

Finally, packaging and labeling requirements are addressed in Section 2-7 of Article XII. Labels must include "manufacturer name, registration number, expiration date, storage conditions, chemical name, trade name, ratio of active substances and purity, and precautions necessary to protect human health and the environment."[57] Further, Section 2-7 includes provisions that:[58]

1. Labels be clearly legible and reflect the true nature of the product;
2. Arabic is the mandatory language on labels, with English optional;
3. Small containers placed inside larger ones are also subject to labeling; and
4. The following information be printed on the label:
   a. Product name;
   b. CAS number;
   c. Ingredients and purity ratio;
   d. Precautionary and hazard statements;
   e. Production and expiry dates;
   f. Storage and temperature instructions;
   g. Country of origin;
   h. Manufacturer name and address; and
   i. Classification and safety measures.

It is suggested that a SDS should be prepared and accompany the label; however, as discussed, the sections and attendant information in the document are not formally required by the GCC.

In 2007, the GCC promulgated GSO 1810:2007 (GSO 1810), "Labeling for Chemical Products." While voluntary in nature, GSO 1810 does present some aspects worth examining. It applies to all chemical products, excluding pharmaceuticals and foods.[59] The product label must reflect the true nature of the product, not be misleading, include relevant Environmental Health and Safety (EHS) information, and be printed in Arabic.[60] It should be noted that if the container has smaller (individual) containers inside it, then these interior containers must be labeled as well.[61]

The specific information which is required on the product labeling according to GSO 1810 is:[62]

- Product name;
- CAS number/scientific name;
- Classification;
- Ingredients and purity;
- Manufacturer's name and address;
- Country of origin;
- P&H statements [Precautionary and Hazard];
- MSDS [it is unclear from GSO 1810 what type of "MSDS" information should be included];
- Treatment and precautionary measures;
- Production and expiry [sic] dates; and
- Handling and storage temperature.

Having examined some of the more general chemical substance regulations within the GCC, attention may now be turned to regulations which manage specific substance categories and specific types of environments and applications. With respect to the former, the 2004 Pesticides Act of Cooperation Council [sic] for the Arab States of the Gulf (Pesticides Act), and its implementing legislation – the Implementing Regulations of Pesticides Act of the Cooperation Council for the Arab States of the Gulf – functions as the overarching GCC pesticide law.[63] As noted earlier, sovereignty resides with each Member State, which may or may not choose to adopt particular regulations. In the case of the Pesticides Act, for example, Kuwait has transposed it into national law as "Law No. 21 of 2009 Approving the Pesticides Act in the Countries of the Cooperation Council for the Arab States of the Gulf."[64]

Although only six (English) pages in length, the Pesticides Act is comprised of 15 Articles, with the manifest aim to regulate production, import, and circulation of pesticides in the Member States of the GCC.[65] Interestingly, Article 2 defines a "pesticide" as "any chemical product that is used mainly for fighting pests and insects and either this pesticide was organic or non-organic, manufactured or natural, or even biological that incorporate[s] microorganism elements."[66] The definition appears to encompass biological pest control methods, something which several other non-GCC pesticide regulations explicitly exclude.

The various incorporations will be discussed in the respective country sections, but in general, the Pesticides Act has several notable sections and requirements:[67]

- Section 4: The import, manufacture, or "exchanging" of any pesticide is not permitted, until or unless a license issued by the relevant authority/authorities is issued according to the rules and procedures stipulated in the Pesticides Act.
- Section 5: The "Minister" is empowered to take the following decisions:
  - To ban the use of importation, sale, or manufacturing of pesticides defined in Section 2 as "restricted";
  - To set forth specific conditions and procedures under which registration of the pesticide product is required, as well as to establish conditions and procedures for the manufacture, import, export, and sale of pesticides in general;
  - To establish the procedures for taking samples, and for analysis of same, in a variety of situations, as well as to set out the procedures to be followed when "re-judging" claims are exercised;
  - To establish the conditions and procedures under which pesticides may be used, as well as how they may be advertised; and
  - To establish the regulations required for pesticides' disposal procedures, in coordination with the relevant authorities.
- Section 9: "Official employees" are granted the authority to enter "places, shops, stores, and corporations that deal with pesticides" to ensure compliance with the Pesticides Act, and empowered to issue fines and other remedies where non-compliance is found. Some specific areas of non-compliance are enumerated in Section 10.
- Section 10: Specific items for which non-compliance may be assessed:
  - Changing, disfiguring, or damaging any of the labels or label details on the packaging;
  - Repackaging pesticide containers without formal approval by the applicable authorities;

- Generating and communicating "publicity and advertising" for pesticides without prior approval from the applicable authority;
- Failing to fully cooperate with Ministry-appointed employees tasked with enforcing the Pesticides Act;
- Importing, selling, or manufacturing pesticides without having received a license, where applicable; and
- Importing, selling, or manufacturing any banned pesticides or those which are determined to have "low quality."

Another area in which the GCC has promulgated regulations to address specific areas of concern may be seen with respect to Occupational Safety and Health (OSH), also known as Health, Safety, and Environmental (HSE) protection. As with many developed nations and/or trading blocs (e.g. MERCOSUR, Andean Community), there exist a wide range of OSH standards, often specific to, or dependent upon, the industry at issue. Indeed:

> a surprising amount of investment has been made in HSE legislation in the GCC. Each state has its own keystone environmental and health & safety laws that underpin all other related laws... Supported by the growth in the oil & gas sector, industry and infrastructure has expanded rapidly, requiring that regulating authorities act accordingly and introduce a full range of HSE legislation covering such things as: hazardous waste disposal, air and water quality standards, equipment safety, hazardous chemicals, asbestos and pesticides.[68]

One theory as to why OSH has become a critical issue in the GCC asserts that the decrease in oil prices (at the time of this publication) has changed the economic mindset of many of the GCC countries. Because of lower market prices, a variety of companies in the region may need to look elsewhere for sources of financing. One such source of potential financing is international banks. Often, as a condition of making such loans, these institutions could impose a wide variety of stringent requirements and conditions on the applicant. OSH regulation and compliance may well fit such. Thus, it is theorized, at least until oil prices recover, that "the need for international investment will continue to play a role in strengthening HSE performance in the GCC."[69]

With respect to OSH regulation, the GCC has developed a wide range of standards. Among these are GSO 209/1994, "Industrial Safety and Health Regulations – Part 3: Occupational Health and Environmental Control" and "Gulf Standard – Industrial Safety and Health Regulations – Part 4: Hazardous Materials – Toxic and Hazardous Substances." GSO 209/1994 relates to the workplace management of a variety of aspects, such as ventilation, noise, radiation, accident prevention (e.g. signs, tags, signaling, and barricades), medical services, and confined or enclosed spaces, describing in considerable detail the measures to be taken to ensure a safe and healthful workplace.[70] The Gulf Standard, by contrast, addresses the industrial safety and health regulations for specific types of equipment and specified means of handling materials. Among these types and means are: conveyors, powered industrial trucks, overhead and gantry cranes, derricks, slings, helicopters, and material hoists.[71]

Additional areas where the GCC has promulgated HSE-related regulations for specific industries and/or aspects are:[72]

- GSO 68/1987 – Industrial Safety and Health Regulations – Equipment – Machinery and Guarding – Part 1: General Requirements;
- GSO 62/1987 – Industrial Safety and Health Regulations: Hazardous Materials – Flammable and Combustible Liquids – Part 1: Tanks, Piping and Accessories;
- GSO 55/1987 – Industrial Safety and Health Regulations: Hazardous Materials – Gases – Part 1: General Requirements;
- GSO 78/1987 – Industrial Safety and Health Regulations – Electrical – Part 4: Illumination; and
- GSO 67/1988 – Industrial Safety and Health Regulations – Buildings – Part 1: Building Facilities.

2

Bahrain

National Overview

The Kingdom of Bahrain (Bahrain) is an "archipelago of 40 islands, with a total area of 757.50 square kilometers. The largest island is the island of Bahrain, which includes the capital (Manama), representing 80.68% of the total area of the islands of the Kingdom."[73] Arabic is the official language, although Persian and English are widely understood and utilized. Bahrain gained its independence from Great Britain on August 15, 1971, although it remained under British protection for four more months, until December 16, 1971.[74]

> Despite its minor role as an oil producer, the country's economy depends heavily on hydrocarbon exports, mostly refined products. Petroleum production and refining account for more than 60% of [the] Kingdom of Bahrain's export receipts and 70% of government revenues. The vast majority of [the] Kingdom of Bahrain's total energy consumption comes from natural gas, with the remainder supplied by oil. Hydrocarbons also provide the foundation for [the] Kingdom of Bahrain's two major industries: refining and aluminum smelting.[75]

Governmental Structure

Bahrain is a constitutional and hereditary monarchy, as a result of having amended its Constitution on February 14, 2002. As such, it recognizes a King as the Head of the State. There are also a Prime Minister, a First Deputy Prime Minister, and four Deputy Prime Ministers. The Cabinet is appointed by the King. "The King presides over the three branches of government (executive, legislative, judicial)."[76] Key among King Hamad bin Isa Al Khalifa's (the current leader's) legislative initiatives are improvements to the political system and related legislation, and institutional machinery within the scope of the reform project focusing on the development of the country's legislation and systems in the political, economic, and social fields.[77]

Bahrain is divided into five municipalities (Muhafazat, Asimah (Capital), Janubiyah (Southern), Muharraq, and Shamaliyah (Northern)).[78] Each of the five has an appointed governor. The legislative branch of the Bahraini government is comprised of a bicameral National Assembly, consisting of the Consultative Council (*Majlis al Shura*) and the Council of Representatives (*Majlis al Nuwab*). Each Council is comprised of

Chemical Regulation in the Middle East, First Edition. Michael S. Wenk.
© 2018 John Wiley & Sons Ltd. Published 2018 by John Wiley & Sons Ltd.

40 members (seats); the Consultative Council members are appointed by the King, while the Council of Representatives members are directly elected by an absolute majority vote.[79] Members serve four-year renewable terms.

The judicial system in the country is a mixed system of Islamic law, English common law, and Egyptian civil, criminal, and commercial codes.[80] "The judiciary of Bahrain is divided into civil law courts and sharia law courts; sharia courts (involving personal status and family law) are further divided into Sunni Muslim and Shia Muslim." The judicial branch consists of multiple layers of courts. The highest court is the Court of Cassation, also known as the Supreme Court of Appeal. It consists of a Chairman and a three-judge panel.[81] Judges at this court are appointed by Royal Decree for a fixed term. One level below is the Constitutional Court, comprised of the President and six members, who each serve nine-year terms. With respect to the sharia courts, Bahrain has established the High Sharia Court of Appeal (comprised of the President and at least one judge).[82]

The Ministry of Information Affairs publishes legislation in the *Official Gazette of Bahrain* (http://www.mia.gov.bh/en/official-gazette/Pages/default.aspx).

Key Chemical Regulatory Agencies

Bahrain receives a not-insubstantial volume of various types of chemicals from around the globe, such as pesticides, cleaning materials, pharmaceutical drugs, and food additives.

> Most of the time, these chemicals enter the country without proper control on the importation and handling, due to the fact that in many cases countries of origin export the chemicals with a lack of information on chemical composition, toxicity, etc. Due to limited capabilities, at present, only a voluntary coordination system for licensing of industrial projects and commercial activities is in place, covering the importation, use, production, storage, and disposal of chemicals.[83]

Within Bahrain, multiple Ministries and related bodies have roles in the management of chemicals. Chief among these is the Supreme Council. The Supreme Council has the overall mandate for environmental protection in the country, which often extends into chemical substance management as well. A unit within the Supreme Council is responsible for managing chemicals through control on their import, export, storage, handling, and disposal. Additionally,

> The Directorate of Environmental Control [(Directorate)] is the lead organization in collaboration with a multistakeholder coordinating committee, comprising other organizations involved in the management of chemicals in [the] Kingdom of Bahrain, namely: Customs Affairs of the Ministry of Interior; Authority of Electricity & Water; Ministry of Industry and Commerce; Agriculture Affairs of the Ministry of Municipalities and Urban Planning; Ministry of Health; Ministry of Interior; Ministry of Labor; Kingdom of Bahrain University; Chemists Society; Engineers Society; NGOs; and other research centers.[84]

The Directorate is tasked to draw up "plans, policies, and mechanisms for the management and protection of the environment and to ensure industry compliance with the environmental standards and regulations for maintaining a safe and healthy environment to achieve sustainable development for the present and future generations."[85] Consistent with this mission, the objectives of the Directorate are:[86]

1. To plan and formulate rules, regulations, policies, standards, and legislation for the control of pollution caused by industries, waste management, and conservation of environmental resources;
2. To prepare, implement, monitor, and ensure compliance with environmental legislation by the users and industries to safeguard human health and the fragile ecosystem of the country;
3. To react and attend to any environmental emergency occurring within the territory of the Kingdom of Bahrain;
4. To control storage, handling, movement, treatment, disposal, import, and export of potential and hazardous substances, wastes and chemicals; and
5. To fulfill the country's local, regional, and international commitments related to the management, monitoring, and control of pollution and environmental resources.

Some of the key mandates of the Directorate, with respect to chemical substance management (in general), are:[87]

1. Prepare and implement rules and regulations for the protection of the environment, safety of people, and waste management;
2. Prepare, participate, and ratify regional and international conventions and protocols on environmental plans for pollution control;
3. Develop and implement plans and procedures for emergency preparedness against pollution disaster and occupational risks;
4. Study applications for opening of new service areas and industrial establishments and ensure compliance with environmental criteria;
5. Conduct studies and field surveys to determine the quantity and quality of waste production and suggest the techniques and methodologies required for its treatment, and to minimize their risks;
6. Develop management and monitoring plans for environmental control of POPs [Persistent Organic Pollutants] sources in the working environment;
7. Establish criteria and procedures for management and monitoring of hazardous waste;
8. Control importation, storage, handling, transportation, transfer, and disposal of hazardous and toxic materials and dangerous equipment, whether chemical, biological, or physical. And control illegal transboundary trade of dangerous goods in the Kingdom of Bahrain's territory and coastal waters;
9. Prepare plans and procedures for reduction, minimization, and recycling of waste;
10. Recommend environmental health criteria and procedures to protect workers against occupational hazards and pollution and to safeguard the public health of citizens and coordinate with the Ministry of Health in this regard; and
11. Prepare programs required for development and training of technical and professional staff.

Continuing to examine Bahraini governmental entities with responsibilities for chemical management, Customs Affairs, under the direction of the Ministry of Interior, supervises the import and export of cargo (including chemicals), and is also responsible for taxation of these cargoes, including chemicals.[88] The General Directorate of Civil Defense holds responsibility for approving and supervising the location of businesses for certain commercial activities, ensuring that sufficient safety measures are in place.[89] The Standardization and Metrication Directorate (SMD), under the supervision of the Ministry of Industry and Commerce, also holds roles pertaining to chemical management. Among these are:

1. To improve productivity, quality, and market access for businesses and industries, protect consumer interests, and enhance safety, health, and environmental conditions for the Kingdom of Bahrain through the appropriate technical regulations and standards; and
2. To develop and enforce technical regulations fully consistent with World Trade Organization (WTO) principles and agreements for imports and locally produced products pertaining to health, safety, and environment conservation.[90]

The Ministry of Municipalities and Urban Planning, through its Agriculture Clinic Division, holds responsibility for issuing licenses for the import of "special chemicals for combating insects and parasites," while the Division of Plant Protection is tasked with issuing pesticide regulations and issuing licenses for the import of agricultural pesticides.[91]

Key Chemical Substance Regulations

Bahrain presently has no national inventory or new chemical notification scheme in force. However, prior to import of a chemical substance, an entity is required to complete a specific chemical application form, and to send it to the Ministry of Industry, Commerce and Tourisms' Environmental Control Directorate, Chemical Safety Unit.[92]

> Bahrain has in place a system to control the importation, use, and misuse of toxic chemicals, as well as for the licensing of industrial establishments, laboratories, and chemical storages. Inspections of work operations and processes related to the use of chemicals are carried out to determine possible sources of industrial and chemical risks, and to establish procedures for importation, handling, transportation, and storage of chemicals.[93]

Perhaps the penultimate regulation pertaining to chemical substance management in Bahrain is Legislative Decree-Law No. 21 of 1996 (Decree-Law No. 21). Decree-Law No. 21 was enacted to establish the Environmental Affairs Agency (EAA) under the Ministry of Housing, Municipalities and Environment in 1996.[94] Decree-Law No. 21, which entered into force on July 13, 1996, formed two key Directorates: The Directorate of Environmental Assessment and Planning and the Directorate of Environmental Control.[95] The Directorate of Environmental Assessment was later upgraded, and its name was changed to the Supreme Council for Environment (SCE), responsible for managing and monitoring of the environmental resources of the country.[96]

The scope of Decree-Law No. 21 is found in Article 1, in one breathtakingly long sentence:[97]

> … to protect the environment from the polluting sources and factors, and put [an] end to its deterioration by drawing up the required plans and policies to preserve it from the harmful effects resulting from activities causing damages to human health, agricultural crops, marine life and wildlife, other natural resources and the climate, and the implementation of such plans, policies, adopt all appropriate procedures and arrangements to put an end to the deterioration of the environment, prevent or combat all types of environmental pollution and limit such pollution for the benefit of the present and future generations through the realization of the consistent development objectives.

Article 2 sets forth the definitions for specific terms used in Decree-Law No. 21. Of specific relevance to chemical substances is the definition of "hazardous materials and waste":

> Any material or waste resulting from industrial operations, chemicals, or radiation and [that] acquires the hazardous characteristics because of its contents of material, material concentrates, chemical reacting [sic], distinguished as toxic, capability to burst and corrode or any other characteristic resulting in danger to human, animal, plant life or the environment whether alone or when it gets in contact with other waste.[98]

Articles 3 and 4 give the "environmental body," the EAA, broad powers to "undertake the issue of resolutions and instructions in all matters relating to the environment and be empowered to exercise all the required authorities and powers" (Article 3), and to coordinate and cooperate with other regulatory bodies, "as necessary," to "prepare drafts of laws, legislation and to issue regulations which ensure the safety, protection, and development of the environment" (Article 4).[99] Notably, "A punishment for a period not exceeding one year and a fine of no more than BD 1,000 or both punishments, shall be inflicted upon everyone who violates the provisions of Article 3."[100] Article 4 also provides a list of 21 aspects for which the EAA has authority to manage, including:[101]

- To observe public and private activities which have a negative effect on the environment;
- To decide and control the measures and limits permitted for the emissions of materials causing pollution of the environment and their concentration levels;
- To determine the standards relating to the import and dealing in chemical and radioactive materials and to supervise their implementation; and
- To lay down the required bases for the proper management of the industrial, health, and domestic waste.

Article 6 of Decree-Law No. 21 codifies the prohibition against environmental degradation, and effectively sets the basis for penalties and sanctions, noting "[i]t shall not be permitted for any person or project to use the environment in any causing environment pollution [sic], contribute to its deterioration activity, cause damage to the natural resources, [or] living beings," while Article 9 prohibits "leaking of materials and agents causing environmental pollution over the maximum limits permitted by the law and the implementation regulations."[102]

With Article 10, we begin to see how Decree-Law No. 21 interacts with other Bahraini regulations, such as Legislative Decree No. 11 (1989) governing pesticides, which will be discussed later. As per Article 10:

> The spraying or using of pesticides or other chemical compounds for agricultural, public health or other purposes shall be prohibited except after considering the conditions, regulations and guarantees determined by the environment body in agreement with the Ministry of Health and the Ministry of Works and Agriculture so as to guarantee that the environment shall not be directly or indirectly affected, whether at present or in future, by the damaging effects of such pesticides or chemical compounds.[103]

Article 13 relates specifically to "activity which may lead to causing pollution or deterioration of the marine environment or the adjacent waters," and notes that "each day of the continuous discharging or prohibited activity shall be considered as a separate violation."[104] Interestingly, this is one of the few areas within Decree-Law No. 21 where a specific mechanism or means of assessing penalties for non-compliance are defined. One must read to almost the very end of the regulation, in Article 29, to see further criteria for penalty assessment. As per that Article:

> Without prejudice to any severer punishment provided for in any other Law, every violation of the provisions of Articles [6 through 20, 24 and 27–28] of this Law shall… [impose] a punishment of imprisonment and a fine no [sic] exceeding BD 50,000… in addition the court is empowered to order the closure of the premises where the work is a source of pollution for period exceeding three months. If the violation is committed again, the court may order the cancellation of the license.[105]

Article 14 mandates the receipt of a permit from the EAA for "dealing in hazardous materials and waste," with the various Ministers, each focusing on their respective areas of authority and responsibility, in coordination with the Minister of Housing, Municipality and Environment, issuing a list of what constitutes "hazardous materials" and "waste."[106] Finally, under Article 15, the "owner of the establishment which produces such hazardous materials as a result of its activity in accordance with the provisions of this Law, shall maintain a register for such waste and the manner of their disposal," with the EAA being granted authority to review such a register to ensure compliance.[107]

The remaining Articles of Decree-Law No. 21 address a variety of requirements for managing pollution and environmental degradation. Of particular note are Article 16, requiring that "projects shall be required to take the necessary precautions and measures to prevent the leakage or emission of materials and polluting agents inside the work area except within the permitted levels"; Article 19, which mandates the management of waste via "waste treatment units"; and Article 24, which sets out the requirement to use "the [best] technology available," as determined by the EAA, to prevent pollution during new projects or major alterations to existing projects.[108] Interestingly, the ensuing Article directs the EEA to consult with the organization at issue to effectively agree on what constitutes such technology in the given situation. Article 29 expressly notes that any sanctions provided for in the Article will not preclude a Bahraini court from requiring the violator to pay all costs associated with environmental remediation relating to the violation, as well as allowing for compensation due to such damages.[109]

Tangentially related to Decree-Law No. 21, in 1999 Bahrain implemented Resolution No. 1 (Resolution). The purpose of the Resolution was to protect the environment by regulating and controlling substances that deplete the ozone layer, which impacts the management of chemical substances in the country. The Resolution is composed of 17 Articles; of note, Article 1 contains terms and definitions, Article 2 lists the substances harmful to the ozone layer, and Article 4 specifies the documents and procedures for obtaining a license from the Environment Board for the importation of substances listed in Article 2.[110]

In 2002, Bahrain promulgated Resolution No. 7 "On the Control of the Importation and Use of Prohibited and Severely Restricted Chemical Substances" (Resolution No. 7), administered by the Environmental Control Directorate of the SCE. Resolution No. 7 defined a specific list of chemical substances, in any physical form, whose importation into or use in the country is either severely restricted or fully banned. As one might expect, banned substances are those whose import, use, and/or production are not permitted at all within the territory of Bahrain. Restricted substances, however, may only be used for the specific use stipulated in the regulations issued by the Authority, the Chemical Safety Group (CSG).[111]

The chemical substances included in Resolution No. 7 are primarily those which have been proven (*Note:* items not <u>proven</u> to cause negative effects, e.g. suspected carcinogens, are not included here) to be highly dangerous to humans, animals, and plants by regional and international organizations, or by research and studies, and may also have a negative impact on the environment due to their toxicity, explosivity, flammability, or corrosivity.[112]

The CSG may grant one or more approvals for the following avenues of restricted substances under Resolution No. 7: import, export, or sale of chemicals and chemical products (although agricultural chemicals, pesticides, detergents, and cosmetics are specifically excluded from this procedure), import, export, or sale of chemicals for asbestos removal, and import, export, or sale of ozone-depleting substances and their alternatives.[113] Notice here how the last category of potential approval by the CSG dovetails with Resolution No. 1. The CSG is granted the authority to permit users and importers of chemicals under the auspices of Ministerial Order No. 5 of 2005, and the permit – if granted – is fairly precise with respect to where the substance(s) may be used. Specifically, "a) the chemical activity should be permitted in the services or industrial areas and b) the location or activity should be at least 60 meters away in all directions from residential areas."[114] Only registered importers of chemicals are potentially permitted to import, use, and sell chemicals to local users.[115]

Resolution No. 4 of 2006 (Resolution No. 4), "On the Management of Hazardous Chemicals," relates to the "proper handling of chemicals, including the productive and industrial processes, import, export, customs clearance, storage, transport and consumption."[116] Resolution No. 4, amended in part by Resolution No. 6 of 2013, applies to all aspects of handling "hazardous chemicals" within Bahrain, with the following exceptions: narcotic and psychotropic drugs, radioactive materials, pharmaceuticals, including medical and veterinary drugs, nutrients, and chemicals used as food additives, explosives, and weapons.[117] Curiously, "hazardous chemicals" are defined as "[a]ny chemical substance with interactive properties inherent in it, in a mixture or a preparation, whether any such substance is in its natural or manufactured form.[118] Thus, by this definition, virtually any substance may be considered "hazardous," as there is some degree of "interactive properties," under certain conditions and in the presence of certain substances and/or compounds, with virtually all substances. As will be discussed later, Article 1 of Resolution No. 4 contains some provisions which relate to SDS and labels, and Article 9 addresses SDS further.

Law No. 15 of 2007 (Law No. 15), under the purview of the Ministry of Health, Pharmacy and Drug Control Directorate, holds responsibility for several of the products and categories exempted under Resolution No. 4. Specifically, Law No. 15 regulates the controlled production, manufacture, import, export, sale, transfer, and possession of narcotic drugs, psychotropic substances, and chemical precursors.[119]

To more easily understand the stages of the chemical substance lifecycle during which Bahraini regulations are applicable, in 2012 the SCE, as part of the "Kingdom of Bahrain National Profile to Assess the National Infrastructure for Chemical Safety," developed the very useful matrix "Overview of Legal Instruments to Manage Chemicals by Use Category."[120]

Category of chemical	Import	Production	Storage	Transport	Distribution/ Marketing	Use/ Handling	Disposal
Pesticides (agricultural, public health, and consumer use)	X	X	X		X	X	X
Fertilizers	X	X	X	X	X	X	X
Industrial Chemicals (used in manufacturing/ processing facilities)	X	X	X	X	X	X	X
Petroleum Chemicals	X	X	X	X	X	X	X
Consumer Chemicals	X	X	X	X	X	X	X
Chemical Wastes	X	X	X	X		X	X

Source: Kingdom of Bahrain National Profile to Assess the National Infrastructure for Chemical Safety 2012. Available at: http://www2.unitar.org/cwm/publications/cw/np/np_pdf/Kingdom of Bahrain_National_Profile_update.pdf.

Pesticide Regulations

Legislative Decree No. 11 of 1989 (Decree No. 11) addresses pesticide management in Bahrain. The purpose of Decree No. 11 is to regulate the production and importation of pesticides.[121] This comprehensive law, covering all synthetic or natural chemical compounds used as pesticides to control any type of pest, addresses such categories as public health vector control pesticides, household pesticides, and professional pest control pesticides, and relates to the use and licensing of agricultural pesticides.[122] All specifications within Decree No. 11 are in line with the World Health Organization (WHO), the Food and Agriculture Organization (FAO), the International Plant Protection Convention (IPPC), and the WTO standards, as well as other international recommendations.[123] Additionally, Legislative Decree 37 of 2005 "Concerning the Approval of Gulf Countries Council (GCC) Unified Pesticides Order" aims to regulate the production, import, and trading of pesticides, while the Fertilizer Law of 2006 relates to the management and regulation of such substances.[124]

Within Bahrain, the Pesticides Registration Committee, part of the Ministry of Works, Municipalities Affairs and Urban Planning's Agricultural Affairs authority, is the competent authority for the registration of both agricultural and public health use pesticides. Ministerial Order No. 110 of 2006 established the Pesticides Registration Committee (Committee) for registration of these two categories of pesticides.[125] The Committee is comprised of representatives of the following bodies:[126]

- Ministry of Municipalities and Urban Planning Affairs, Directorate for Agriculture (chair);
- Ministry of Municipalities Affairs and Urban Planning, Agriculture Planning and Development Authority;
- Public Commission for the Protection of Marine Resources, Environment Wildlife, for Environment Control Directorate;
- Ministry of Interior, The General Directorate of Civil Defense – Protection and Safety Directorate;
- Ministry of Industry and Commerce, Standards and Metrology Directorate;
- Ministry of Interior, Customs Affairs;
- Ministry of Health, Public Health Directorate, Environment Health Section; and
- Ministry of Interior, General Directorate of Investigation and Criminal Evidence.

For the Committee to fully consider the registration of agricultural or public health pesticides, a copy of the Commercial Register of Activity for pesticides, as well as a copy of the Commercial Pesticide Registry (CPR) for the owner of the company or organization, must be provided. Additionally, the following information shall be submitted as part of the application process:[127]

1. A letter informing the [C]ommittee that they wish to register a pesticide;
2. The required registration form;
3. All data required for pesticide registration [see below], as well as a certificate of registration in the country of origin (in either Arabic or English) [emphasis the author's], which is consularized by the Embassy in the country of origin;
4. The applicant must provide a certificate from the manufacturer which demonstrates product quality, as well as results of an analysis which proves the product is compatible, qualitatively and quantitatively, "to the substrates and impurities of the pesticide";
5. Methods of analyzing the components of the pesticide and its impurities in concert with the methods adopted by the FAO and the WHO;
6. Evaluation studies, especially those regarding carcinogenic effects and other negative health effects;

7. Any technical publications regarding the pesticide, translated into Arabic <u>and</u> English [emphasis the author's]. This should further include the specifications of the pesticide and its application, as per the specifications of the FAO and the WHO, as well as other necessary information that the Committee may request;

8. The Authority may request testing of the compound, with the resultant documentation to be provided in Arabic <u>and</u> English [emphasis the author's] on the prescribed form;

9. Toxicity studies which are available relating to toxicity to the environment and to mammals;

10. Information and protective chemical properties of the composite;

11. Information relating to the storage and disposal of used containers of the pesticide and any excess thereof, as well as any other information deemed necessary for the registry;

12. A copy of the proposed labeling which includes information on the pesticide, as well as the proposed trade name of the pesticide that will be used in the country, with copies translated into Arabic <u>and</u> English [emphasis the author's];

13. Four original samples of the pesticide, each of 1 l/kg, as well as an additional one, to be used to conduct field tests; and

14. Samples of the active ingredient(s), along with a sample of the original formulation of a pesticide, to confirm the quality tests.

Readers will note that only the certificate of registration in the country of origin may be presented either in Arabic or in English. Where other data points have specific requirements with respect to language, they are to be provided in both Arabic as well as English. Based on the results of laboratory and field testing, the Committee will decide whether to approve the registration. If the Committee refuses to register the pesticide, then the applicant is notified and provided with the reasons for rejection. The registration form will be referred to the Committee in their first meeting after all the required documents are received.[128]

Other key sections of Decree No. 11 are:[129]

- Labeling of pesticides: Point 2 of Article 6 (Legislative Decree No. 11 of 1989 governing pesticides);
- Registration of pesticides: Article 8 (Legislative Decree No. 11 of 1989 governing pesticides);
- Licenses for pesticides: Articles 9 to 12 (Legislative Decree No. 11 of 1989 governing pesticides);
- Inspections regarding pesticides: Article 13 (Legislative Decree No. 11 of 1989 governing pesticides);
- Licenses for setting up or managing an industrial establishment: Articles 1 to 15 (Legislative Decree No. 6 of 1984 governing the organization of industry);
- Registration of industrial establishments: Articles 16 to 27 (Legislative Decree No. 6 of 1984 governing the organization of industry);
- Inspection of industrial establishments: Article 28 (Legislative Decree No. 6 of 1984 governing the organization of industry);
- Use of pesticides: Article 10 (Legislation Decree No. 21 of 1996 concerning the environment);

- Permits for discharge and storage of wastes: Articles 13 and 18 (Legislation Decree No. 21 of 1996 concerning the environment); and
- Production, handling, and disposal of dangerous materials: Articles 14 and 15 (Legislation Decree No. 21 of 1996 concerning the environment).

Occupational Safety and Health Regulations

Bahrain has issued a variety of legislation as well as Ministerial Orders pertaining to OSH over the past four decades. As will be seen with several other countries, many of the OSH regulations which have been promulgated tend to address specific actions and/or specific industries, as opposed to taking the form of an overarching and all-encompassing OSH regulation. Among the key Bahraini OSH regulations are:

- The Bahrain Labor Law (2012). "Under the umbrella of this law, there are 20 Ministerial Orders [27 at press time] issued to regulate OSH at workplaces. These orders [cover] the management of OSH on establishment level, operations safety, health and welfare, and fire prevention";[130]
- Ministerial Order No. 25 (1977), "Precautionary Measures for Protection of Workers in Workplaces"; and
- Ministerial Order No. 31 (1977), which addresses "the protection of workers from the hazards of highly flammable liquids and liquefied petroleum gases."[131] This Order relates to the use, storage, and transport of certain highly flammable substances in both indoor and outdoor workplaces. It also sets out the duties of employers and employees.[132]

The Bahraini Ministry of Labor and Social Affairs established the "Occupational Safety Section" (Section) as part of the Labor Directorate.[133] The Section holds authority for OSH in the country, which includes the safe handling of chemicals, as set out in the 1976 Labor Act for the Civil Sector (Labor Act). The Section carries out its inspections on a random basis. Perhaps uniquely, "there are no specialized campaigns based on the level of existing hazards or on the basis of past history."[134] Normally, inspectors review the existing hazards and the level of risk at a given establishment and will, based on their experience in this regard, determine if the measures taken by the organization to protect its workers are adequate.[135] Based on the observations and findings from such on-site inspections, Section inspectors will determine the actions to be taken in order to ensure compliance with OSH standards and regulations.[136] These may take the form of a verbal notice to the employer, an "Improvement Notice" (a written document showing the violations found and warning the organization that it is out of compliance), a "Prosecution Notice" (a document to be transferred to the judicial authorities), or an "order for closure of the establishment," either for the whole or any part(s).[137]

The central purpose of the Labor Act is to "regulate the conditions of employment in factories and other places of work with regard to safety, health, and welfare of persons employed."[138] The Act covers all hazards that are found in workplaces, including a wide range of places of employment where chemicals are formulated, used, and packaged for trade. The Section enforces the regulations of the Bahrain Labor Law, as well as 27 Ministerial Orders related to OSH.[139] The Labor Act primarily functions as a vehicle to allow the authority to periodically inspect private sector establishments, and to take legal action in situations where conditions in such establishments violate the safety and health regulations. In addition, the Labor Act permits the Ministry of Labor to investigate occupational safety and health complaints made by workers, as well as industrial accidents.

The Ministry of Labor is also a member of the Bahraini National Supreme Health Committee, created by Ministerial Order No. 4 (1998). The Committee, consisting of 14 members and comprising all of the major OSH stakeholders in Bahrain, has the manifest purpose to advise the Ministry on the future direction of OSH in the country. The Committee:

> in cooperation with the Occupational Safety Section in [the] Ministry of Labor, carries out research and studies in the field of occupational health, organizes conferences, seminars, and workshops on health and safety, and supports occupational safety and health. The Committee is also responsible for the ratification of the international conventions on health and safety.[140]

Waste Regulations

> Waste has become one of the main challenges for sustainable development in the Kingdom of Bahrain. The country is currently struggling to manage wastes from multiple sources… largely due to the accelerated increase in waste volume as problems related to limited geographical area, scarcity of safe waste-disposal sites, and lack of environmentally appropriate technologies for waste handling and treatment.[141]

Key among the Bahraini waste management regulations is 2006's Ministerial Order No. 3 (Order No. 3), "Concerning the Management of Hazardous Wastes," which was implemented by Resolution No. 3 ("Hazardous Waste Management") of the same year. Ministerial Order No. 7 of 2002 is also a waste-related statute; however, as it contains information relating to MSDSs, it will be discussed in the "Safety Data Sheets and Labels" section.

Order No. 3 established monitoring and control systems to manage the generation, storage, transport, and treatment of hazardous waste. Regarding hazardous medical waste specifically, the Kingdom has promulgated Bahrain-enacted Ministerial Order No. 1 "Concerning Management of Hazardous Medical Waste" in 2001. As the name implies, Ministerial Order No. 1 sets out rules for the control and supervision of the production, storage, transport, treatment, and disposal of waste, to control such waste and prevent its harmful effects on health and the environment, until the point at which the substances are disposed of in a proper manner.[142]

Order No. 3 aimed to establish a management system to "control the production, storage, transport, treatment, and disposal of hazardous wastes and their export for treatment in order to control these wastes and prevent the spread of their harmful effects."[143] Order No. 3, under the auspices of the SCE (the competent authority) is applicable to "every product, conveyor or treatment unit for hazardous waste or disposed of whenever its activity is related to or wholly or partially related to hazardous waste… [but] excludes non-hazardous industrial wastes such as commercial, household, fertilizer, agricultural, and animal."[144]

Article 1 of Order No. 3 proffers a comprehensive definition of "hazardous waste":

> Any solid, semi-solid or liquid substances containing gaseous residues or a group of waste compounds that pose a risk or potential risk to public health, the environment or wildlife due to their physical, chemical or biological properties, concentrations or characteristics when administered in a manner [that is] environmentally sound and include the following:[145]
>
> A) All wastes containing the properties listed in Annex IV of this Resolution shall include chemical residues and know-how as non-usable chemical products or non-conforming products or residues of containers of residue materials or residues that belong to one of the categories listed in Annex III;
> B) All wastes belonging to one of the categories listed in Annex III and shall be characterized by any of the properties listed in Annex IV or if they are a mixture of hazardous waste with other substances;
> C) Any residue that exceeds the concentration of the measures listed in Annex V after the Toxicity Characteristic Leaching Procedure (TCLP);
> D) All hazardous wastes mentioned in Annex VI of this Resolution;
> E) Any other wastes classified by the competent department as hazardous waste.

Hazardous wastes are differentiated from "hazardous substances" within the same Article as "[a]ny solid, semi-solid, liquid or gaseous substance containing different amounts and concentrations that cause a hazard to public health or the environment if these substances are not properly managed."[146]

It should be noted that Article 4 of Order No. 3 gives the SCE the express ability to reclassify and characterize the hazardous (and non-hazardous) waste cited in the Annexes to the decision, as well as to amend the requirements as well as the controls needed to implement the foregoing. As such, entities should carefully follow the legislation to ensure they are up-to-date on the most recent changes.

Perhaps interestingly, the TCLP concentrations referenced in item C) above are identical in both value and substance to those included in the United States' Resource Conservation and Recovery Act (RCRA) (cf. 40 C.F.R. § 261.24), save for the fact that RCRA specifically delineates the *ortho-*, *meta-*, and *para-* forms of Cresol, as well as the Total, while Order No. 3 only incorporates the Total value.

Article 5 of Order No. 3 introduces a concept somewhat unique to the region: waste minimization. The Article requires the generators of any hazardous waste product to prevent, or at least reduce, its generation via the equipment used and the selection of clean technology methods. The latter concept is suggested to be achieved by the substitution of materials which are less harmful to the environment, wildlife, and public health.[147]

The ensuing several Articles (Articles 6 through 11) lay out the procedural means of managing hazardous waste:[148]

• Article 6: Requirement for each hazardous waste product to complete a "waste data form" approved by the competent department;

- Article 7: Permits the competent department to request laboratory analysis of the hazardous waste as desired;
- Article 8: Requires the generator to classify and characterize the hazardous waste when "sorted, collected, stored, transported, processed, or disposed of" in accordance with the regulation;
- Article 9: Requires the storage of hazardous wastes in drums or other specialty containers, and to temporarily store them in the facility away from workers and where they may potentially harm the environment;
- Article 10: The generator may choose to establish and operate hazardous waste treatment units in the facility, with prior written approval from the competent department; and
- Article 11: This Article addresses how hazardous waste must be transported once it leaves the facility, including container specifications, applicable forms, compliant shipping means, and so forth.

Articles 12 and 13 relate to the proper means of disposal of such waste, while Articles 14 through 16 address the requirements (e.g. licensing, training/experience of personnel, etc.) for the management of hazardous waste-treatment facilities, while Articles 17 through 19 address transporter requirements. Note that these Articles are applicable to the entity which actively transports the waste to a licensed treatment or disposal facility, while Article 11 focuses <u>solely</u> on requirements for the generator.

Articles 20 through 28 prescribe the administrative requirements for the official establishment of the hazardous waste treatment unit and disposal sites, as well as the technical standards applicable. The ensuing Article expressly directs that any "natural or legal person" is prohibited from importing hazardous waste, either for treatment or disposal, into Bahrain, and goes so far as to prohibit the "entry or passage of such" into the country, "for any purpose or cause."[149] Article 30 does, however, make an express exception for ships carrying hazardous waste to pass through "the territorial sea of the Kingdom of Bahrain and its adjacent area," provided the concerned authority and the competent department give authorization.[150] Additionally, as per Article 31, no "natural or legal person" is permitted to export hazardous waste from the country, unless they possess a license issued by the competent department. The license application requirements are further defined in Articles 32 through 34.

On January 7, 2015, the Prime Minister issued Prime Ministerial Edict No. (2) of 2015 on the formation of the Safety and Occupational Health Council.[151,152] Comprised of 16 members representing a variety of Ministries, "[t]he Council shall be responsible for formulating and following up the implementation of the general policy in the field of occupational safety and health, and securing the work environment."[153] Each of the 16 members is appointed to a three-year term, renewable for a similar term if desired.

Meeting at least once every three months, the Council holds responsibility for formulating and following up on the implementation of the general policy on occupational safety and health, and securing the work environment through the following 11 duties and functions presented in Article (4):[154]

1. Proposing a national policy and system for occupational safety and health and securing the work environment;
2. Proposing and review of the development of national legislations and standards related to the protection and safeguarding of the safety and health of workers and

the work environment in light of practical changes, scientific advances, developments related to economic and social development programs, and related ratified Arab and international agreements pertaining to occupational safety and health and work environment;

3. To give its opinion on proposed legislations that are related to occupational safety and health and work environment;

4. Conducting studies and research related to occupational safety and health and work environment;

5. Studying Arab and international agreements and recommendations related to occupational safety and health and work environment;

6. Seeking to benefit from the exchanging of experiences with firms, societies, committees, and organizations working in the field of occupational safety and health and work environment;

7. Developing and proposing plans related to training and grooming staff cadres specialized in the field of occupational safety and health and work environment in line with practical and scientific needs, **as well as determining the qualifications required for occupational safety and health professions** [emphasis the author's];

8. Promoting preventive awareness in the field of occupational safety and health and work environment, through TV, radio, and press media, and the organizing of specialized exhibitions, conferences, and seminars in coordination with governmental bodies, major companies, and civil society organizations;

9. Consulting, coordinating, and cooperating with other concerned bodies in proposing laws, regulations, and measures related to occupational safety and health and work environment;

10. Coordinating directly with the occupational safety and health inspection function at the Ministry of Labor in the area of the implementation of occupational safety and health laws and Ministerial Orders; and

11. Providing consultation services in the area of occupational safety and health to the occupational safety and health inspection function at the Ministry of Labor.

Interestingly, Article (8) provides the Council with the authority – should it take such a decision – to "set up occupational safety and health subcommittees in related economic and industrial sectors as it may determine, provided that every committee shall include in its membership two representatives from the concerned parties."[155] Thus, the Council has a direct means of engaging with a variety of other concerned stakeholder groups in order to facilitate discussions, and such discussions must include members of the stakeholder groups, thereby increasing the level of involvement external to the government.

Work in this Council is ongoing.

Safety Data Sheets and Labels

Ministerial Order No. 7 of 2002 (Ministerial Order 7) sets out some requirements for MSDSs (in this Order, Bahrain uses the older term of Material Safety Data Sheets instead of the newer term Safety Data Sheets) for chemicals and chemical products in Bahrain, although it is largely vague with regard to specific requirements. As per Chapter II,

Article 2 (interestingly, the "Objective and Scope of the Order" is Chapter II, while the "Definitions" are Chapter I. This is the reverse of how many regulations globally present their texts – beginning with the scope of the regulation, and then following up with the definitions of the terms therein), the "Objective and Scope of the Order" is:

> This Order is intended to establish a monitoring and control system suitable for the processes involving hazardous wastes production, storage, transportation, treatment and disposal, a system also designed to manage such wastes and prevent the spread of their detrimental effects to health and the environment until they are disposed of in an environmentally friendly manner. The Order also aims at encouraging and developing treatment processes for this type of wastes in a way conducive to the protection of the public health and prevention of environmental pollution.[156]

Article 3 of Chapter II directs "[t]he Order shall apply to every producer, carrier, and waste treatment or disposal unit once their activity relates totally or partially to hazardous wastes. The Order excludes industrial, commercial, home, inert, agro, and animal wastes…"[157] Thus, any amount of hazardous waste activity makes the producer and so on subject to the regulation – there is no *de minimis* exception set forth herein.

Chapter I, Article 1, Section 15 establishes a fairly well-known definition of an MSDS: "[t]he waste related form which embodies all data such as physical and chemical properties, methods of handling, storage, transportation, and disposal and safety procedures."[158] Remarkably, the definition does not specify whether a "waste" must be hazardous or non-hazardous to necessitate an MSDS, and also appears only to be applicable to "waste." Article 7, however, directs "Each producer of hazardous wastes shall complete the MSDS form in accordance with the format approved by the Authority."[159] As such, the "person" or "producer" (as defined in Chapter I) must infer when the creation of an MSDS is required. Of further interest is the definition in Chapter I, Article 1, Section 9 of a "producer": "Any installation whose activity will **probably** or **could** result in generating waste" [emphasis the author's].[160] This definition is counter to that which appears in many other waste-related statutes globally, where a "producer" is generally defined and/or understood to be an entity that actually <u>creates</u> waste(s), either intentionally or indirectly. A final point of interest lies in the level of segmentation Bahrain uses to define "wastes." Again, many other global waste regulations simply differentiate between "hazardous" and "non-hazardous" wastes, although some legislation does go a bit further and employ categories such as "industrial" and "commercial" waste. Ministerial Order 7 segments among "hazardous wastes," "domestic wastes," "inert wastes," "industrial and commercial wastes," "agricultural wastes," and "animal wastes," all within the same regulation.

As noted earlier, Articles 1 and 9 of Resolution No. 4 additionally have an MSDS requirement. Therein, it is specified that the language of the document must be in <u>both</u> Arabic and English, although no guidance as to format or sections is provided.[161] MSDS for chemicals and chemical products need to be submitted to the CSG.[162]

Labeling requirements for chemical substances are set out in Article 1 of Resolution No. 4. As per this Article, there are two types of labels: "hazard labels," which are square-shaped cards attached at a 45-degree angle to the packaging container of hazardous chemicals, and "handling labels," informative labels which are rectangular in shape and are required either alone or in conjunction with the hazard label(s) for hazardous materials.[163] Article 9 of Resolution No. 4 sets forth the items which are to be included on a label:[164]

- Manufacturer's name;
- Emergency contact details;
- Registration number in the country of origin (e.g. EU registration number under the Registration, Evaluation, Authorization and Restriction of Chemicals (REACH));
- Production and expiry date of packaged chemical(s);
- Name of chemical, including trade name;
- Active substance(s);
- Purity ratio and impurity specifications;
- Necessary precautions to be taken to avoid unintended contact with humans and the environment; and
- Guidelines on treatment in case of poisoning.

3

Egypt

National Overview

The Arab Republic of Egypt (Egypt) is a presidential republic consisting of 27 governorates (*muhafazat*), comprising a total area of about 1,000,000 square kilometers, of which only approximately 36,000 square kilometers (3.6%) are populated.[165,166] It is the most populous country in the Arab world. Approximately 95% of the population lives within 20 km of the Nile River and its delta, and most areas of the country remain sparsely populated or uninhabited.[167]

Great Britain seized control of Egypt's government in 1882, but gave it partial independence from the United Kingdom in 1922, and returned it to full sovereignty in 1953.[168]

As of June 2016, the country had an estimated population in excess of 94,000,000 people.[169] Arabic is the official language; however, English and French are widely understood by the educated classes.[170]

Governmental Structure

Egypt's Constitution sets out a parliamentary system of government, within which the President is recognized as the Chief of State, and the Prime Minister as the head of the government.[171] The President is elected by absolute majority popular vote for a four-year term, with a second term possible. The Prime Minister is appointed by the President.[172]

The Legislative branch is comprised of a unicameral 596-seat House of Representatives (*Majlis Al-Nowaab*).[173] Of these 596 members, 448 members are directly elected by an individual candidacy system.[174] These, and the remaining members, are elected for a five-year term.[175]

> The President and the members of the People's Assembly may propose legislation, although legislation proposed by Assembly members must be referred first to a special committee to assess its suitability for consideration by the Assembly. Acceptable draft laws are referred to a committee of the Assembly for a report. Once legislation is ratified by the People's Assembly, it is submitted to the President, who must approve or veto the legislation within 30 days. A presidential veto can be overruled by a two-thirds majority of the People's Assembly. If the President neither approves nor vetoes the legislation within 30 days, the legislation becomes law.[176]

Chemical Regulation in the Middle East, First Edition. Michael S. Wenk.
© 2018 John Wiley & Sons Ltd. Published 2018 by John Wiley & Sons Ltd.

Egyptian law has its roots in the Islamic legal tradition. The country remains a secular Arab nation, although it has elements of both a common and a civil law system.[177] It has a mixed legal system based on Napoleonic civil and penal law, Islamic religious law, and vestiges of colonial-era laws. While legislation is the primary source of law, the writings of the Muslim *Hanafi* school are applicable to matters of family law, and non-Muslim citizens may fall under Christian or Jewish law if applicable.[178] "… Hanafi is one of four 'schools of law' and considered the oldest and most liberal school of law. Hanafi is one of the four schools of thought… of religious jurisprudence… within Sunni Islam… [making] considerable use of reason or opinion in legal decisions."[179]

The Judicial branch is composed of multiple levels of court. The highest court, the Supreme Constitutional Court (SCC), consists of the Court President and 10 justices.[180] The Court of Cassation (CC) consists of the Court President and 550 judges, organized in circuits with cases heard by panels of 5 judges.[181] As per the 2014 constitution, all judges and justices are selected by the Supreme Judiciary Council and appointed by the President of the Republic, with lifetime appointments.[182]

Egypt publishes its legislative notifications in its Official Gazette, the *Official Journal of the Arab Republic of Egypt* (http://alamiria.com/). As with many other countries in the region, the original text of all regulations is in Arabic, and such are the only official versions.

Key Chemical Regulatory Agencies

Over the past several decades, Egypt has moved rapidly toward industrial and economic development. Part and parcel of that development has been an increase in the number of industrial chemical firms, and the overall rise of the chemical industry. Historically, and to a great degree still, Egypt imports almost all the chemicals it needs.[183]

Related to this increase, a variety of environmental and other challenges have taken root in the country. Indeed, "different chemical substances are being used in… [a variety of applications]… in Egypt. Currently, an exceedingly large number of chemicals is [sic] imported, manufactured, marketed, transported, stored, and disposed of, thus creating huge benefits, but also health and environmental risks."[184]

Recognizing a clear need to manage chemical substances in a safe manner throughout the country, in 1994 the Minister of State of Environmental Affairs (MSEA) was charged with implementing the newly passed Law Number 4, also known as the Environment Law, Through the enabling legislation of Prime Minister's Decree No. 338 of 1995, the Egyptian Environmental Affairs Agency (EEAA), under the direction of the MSEA, was granted authority for the management of chemicals in the nation.[185] The EEAA replaced the agency established by Presidential Decree No. 631 (1982), in all its rights and obligations.[186]

At the central level, the EEAA represents the executive arm of the Minister of State, which itself was created in June of 1997, under the auspices of Presidential Decree No. 275/1997. The EEAA's principal functions are to formulate environmental policies, develop and undertake plans for environmental protection, and promote environmental regulations between Egypt and regional and national organizations.[187] Additionally, the EEAA is the national body responsible for drafting Egypt's policy(ies) with regard to environmental and natural resource protection, endorsing plans and programs to meet the policy goals, and the promulgation of legislation relating to these topics.[188]

Among the key roles of the EEAA are:[189]

- Preparing draft legislation and decrees related to the fulfillment of its objectives;
- Preparing state-of-the-environment studies and formulating the national plan for environmental protection and related projects;
- Setting the rates and proportions required for the permissible limits of pollutants;
- Periodically collecting national and international data on the actual state of the environment and recording possible changes;
- Setting the principles and procedures for mandatory Environmental Impact Assessment (EIA) of projects; and
- Participating in the preparation and implementation of the national and international environmental monitoring programs and employing data and information gained thereof.

The Board of Directors of the EEAA is comprised of the Minister of Environmental Affairs (Chairman), the Executive Head of the EEAA (Vice Chairman), and the following considerable number of additional representatives:[190]

- One representative at least from each of the following six Ministries to be selected by the concerned Ministers: Agriculture, Animal and Fisheries Resources and Agrarian Reform, Public Works and Water Resources, Transportation and Communications, Industry, Interior, and Health;
- Two experts in the field of the [e]nvironment selected by the competent Minister for Environmental Affairs upon a proposal by the EEAA Executive Head;
- Three representatives from non-governmental organizations concerned with the environment to be selected from among candidates of these organizations in agreement with the competent Minister for Environmental Affairs;
- One of the senior employees of the EEAA selected by the competent Minister for Environmental Affairs upon a proposal by the EEAA's Executive Head;
- The Head of the concerned Legal Counsel Department from the State Council;
- Three representatives from the Public Business Sector selected by the competent Minister for Environmental Affairs from among the nominees of the Executive head of the EEAA; and
- Two representatives from universities and scientific research centers, selected by the competent Minister for Environmental Affairs from among the nominees of these bodies.

The EEAA created the National Hazardous Substances Information and Management System (EHSIMS), an online system initiative to facilitate the procedure for obtaining licenses for handling hazardous substances. Although the EHSIMS project is now completed, its initial goals were to promote the safe handling of hazardous substances, as well as establishing and maintaining a unified permitting form and full capability for tracing hazardous substances.[191] The EHSIMS included a database containing 1,817 chemicals, an information network connected with six line Ministries, a unified permitting form to eliminate the duplication of some hazardous substances in different lists, a hazardous substances website to raise public awareness for all the stakeholders, a classification service for hazardous substances, and a unified list of hazardous substances classified as banned, subject to prior approval, or which may enter without restriction.[192]

Presently, there are seven Ministries involved in the issuance of permits relating to chemical substances:[193]

- Ministry of Agriculture and Land Reclamation – hazardous agricultural substances and waste (including pesticides and fertilizers);
- Ministry of Industry (MOI) – hazardous industrial substances and waste;
- Ministry of Health and Population – hazardous pharmaceutical, hospital, and laboratory substances and waste, and domestic insecticides;
- Ministry of Petroleum – hazardous petroleum substances and waste;
- Ministry of Electricity (Nuclear Energy Authority) – hazardous substances and waste from which ionizing radiation is emitted;
- Ministry of Interior – hazardous flammable and explosive substances and waste; and
- Ministry of Environmental Affairs – all other hazardous substances and waste.

Two of the Ministries identified above play a larger role than the others in the regulation of chemical substances in Egypt, and therefore should be noted in more detail. The Ministry of Health and Population (MOHP) established a unit for chemical safety in 1995, to survey chemical incidents and to develop public awareness of the problems arising from chemicals.[194] In addition, the MOHP established a comprehensive program for poison control and chemical emergencies, while also directing the National Committee for Pharmaceutical and Medication Programming to regulate the household use of chemical products.[195] The Ministry of Agriculture and Land Reclamation (MALR) regulates chemical fertilizers and pesticides, controlling the importation and use of these items through different departments to prevent plant diseases and pests.[196]

Key Chemical Substance Regulations

Prior to the early 1980s, environmentally related regulations had been implemented somewhat haphazardly across a variety of Ministries, including the MOHP, the MALR, and the Ministry of Water Resources and Irrigation (MWRI), based on their respective jurisdictions and issues.[197]

Egypt presently has neither an existing chemical substances inventory nor a new chemicals notification scheme.[198] However, as early as 2006, Egypt was "actively engaged in efforts toward the sound management of chemicals, including efforts to reduce POPs."[199] Indeed, "Egypt has issued a large number of environmental legislations governing importing, manufacturing, trade, usage of chemicals [sic] covering different areas."[200] Furthermore, Egypt developed a National Implementation Plan (NIP) to strengthen the national capacity and to enhance knowledge and understanding amongst multiple stakeholders with respect to the regulation of chemical substances.

Among the earliest chemical management regulations in Egypt was Law No. 21 of 1958, also known as the Industry Law, under the auspices of the MOI. Although Law No. 21 ostensibly concerned the promotion of industry within Egypt, it also set out the regulations for the handling and importing of chemicals used by the group.[201] Interestingly, as the following table published by the EEAA illustrates, of the 29 chemical substance-related regulations issued in Egypt, all are considered to be only "fair" in their level of enforcement on a scale ranging from "Effective" to "Weak":[202]

References to existing legal instruments which address the management of chemicals[203]

Legal Instrument (Type, Reference, Year)	Responsible Ministries or Bodies	Chemical Use Categories Covered	Objectives of Legislation	Enforcement Ranking*
Law No. 4 of 1994	MSEA [Ministry of State for Environmental Affairs]	Industrial chemicals, Agricultural chemicals (pesticides, fertilizers), Petroleum products, Consumer chemicals, and Chemical waste	Environmental protection and pollution control in Egypt	(2)
Decree No. 338 of 1995	MSEA	Industrial chemicals, Agricultural chemicals (pesticides, fertilizers), Petroleum products, Consumer chemicals, and Chemical waste	Executive Regulations for Law No. 4/1994	(2)
Decree No. 55 of 1983	MOMI [Ministry of Manpower and Immigration]	All chemicals used in the industrial field	Regulate and control use, handling, and storage of chemicals and conditions required for industrial safety and health in the workplace	(2)
Decree No. 116 of 1991	MOMI	All chemicals used in industrial field	Strengthening the facilities with training for directors and workers	(2)
Decree No. 60 of 1986	MOA [Ministry of Agriculture]	Pesticides	Regulates and controls use of restricted compounds	(2)
Decree No. 258 of 1990	MOA	Fertilizers	Regulates and controls the import of fertilizers	(2)
Decree No. 7330 of 1994	MOIn [Ministry of Interior]	Explosives	Determination of substances that are considered as explosives	(2)
Decree No. 18039 of 1995	MOIn	Explosives	Issue of license for import and use of explosives	(2)
Decree No. 499 of 1995	MOI	Poisonous and non-poisonous substances in industry	Control of handling the poisonous and non-poisonous substances in industry	(2)
Labor Law No. 137/1981	MOMI	Industrial chemicals	Labor and industrial safety protection of industrial environment	(2)

References to existing legal instruments which address the management of chemicals[203]

Legal Instrument (Type, Reference, Year)	Responsible Ministries or Bodies	Chemical Use Categories Covered	Objectives of Legislation	Enforcement Ranking*
Law No. 21/1958	MOI	Industrial chemicals	Rules regulating industry and production, handling and importing of chemicals	(2)
Decree No. 480/1971	MOHP [Ministry of Health and Population]	Industrial chemicals	Air pollution criteria for industrial establishment	(2)
Agriculture Law No. 53/1966	MOA	Agricultural chemicals	Rules regulate production, import, and use of pesticides and fertilizers	(2)
Decree No. 50/1967	MOA	Pesticides	Toxic properties of pesticides and procedures for recording them	(2)
Decree No. 590/1984	MOA	Fertilizers	Rules regulate production, import, and use of fertilizers	(2)
Decree No. 278/1988	MOA	Veterinary insecticides	Regulates importing of veterinary insecticides	(2)
Decree No. 874/1996	MOA	Pesticides	Regulates importing, handling, and use of pesticides	(2)
Law No. 59/1960	MOHP	Ionized radiation's [sic]	Regulates the work with ionized radiations and protection from their danger	(2)
Decree No. 630/1962	MOHP	Ionized radiation's [sic]	Executive Regulations for Law No. 59/1960	(2)
Decree No. 348/1996	MOHP	Banned insecticides	A list of insecticides not allowed to be imported, produced, or used	[not provided]
Decree No. 392/1964	MOHUUC [Ministry of Housing, Utilities, and Urban Communities]	Explosives	Determinations for conditions for explosive warehousing	(2)
Decree No. 138/1958	MOI	Industrial chemicals	Regulates importing, handling, and use of industrial chemicals	(2)
Law No. 113/1962	MOHP	Pharmaceutical chemicals	Regulates importing, manufacturing, and trade of pharmaceutical chemicals	(2)

References to existing legal instruments which address the management of chemicals[203]

Legal Instrument (Type, Reference, Year)	Responsible Ministries or Bodies	Chemical Use Categories Covered	Objectives of Legislation	Enforcement Ranking*
Decree No. 413/1996	MOHP	Hazardous chemicals & wastes	How to get license for handling of hazardous chemicals and wastes	(2)
Decree No. 8/1990	MOHP	Natural and artificial colors	Determination of natural and artificial colors allowed to be used in food industry	(2)
Decree No. 673/1999	MOP	Petroleum hazardous chemicals	A list of hazardous chemicals for the Ministry of Petroleum	(2)
Decree No. 82/1996	MOHP	Hazardous chemicals (for health)	A list of hazardous chemicals for the Ministry of Health	(2)
Decree No. 55/1996	MOT [Ministry of Transportation]	Banned chemicals	A list of chemicals not allowed to be imported, produced, or used	(2)
Decree No. 151/1999	MOI	Hazardous industrial chemicals	A list of hazardous chemicals for the Ministry of Industry which cannot be used without license	(2)

*Effective (1), fair (2), or weak (3) enforcement.
Source: Egyptian Environmental Affairs Agency.

Of particular note from the table above are several pieces of legislation which effectively defined chemical substance management and regulation in Egypt. Law No. 499/1995 instructs that the Ministry of Industry is the responsible Ministry for handling of poisonous and non-poisonous chemicals used in Egyptian industry.[204] Accordingly, the Ministry of Industry holds responsibility for issuing rules and regulations as to the importation of, and trade in, such substances.[205] Relatedly, Decree No. 471/1995 mandates that the Ministry of Industry must be informed of "any activity concerning trade in poisonous or non-poisonous substances including the name of the shop owner, the number of this license, and the kind of trade."[206]

Decree No. 138/1958, as amended by Decree No. 91/1959, directs:[207]

1. For trading in poisonous or non-poisonous chemicals used in industry, a license must be issued from the Industrial Control Authority (ICA);
2. However, an entity may not hold such a license if they presently own a pharmaceutical enterprise (*Note:* this prohibition is due to the fact that "pharmaceutical affairs," including "establishments, personnel, products and ingredients," are subject to the jurisdiction of the Ministry of Health];
3. This license is personal and cannot be transferred or inherited;

4. Poisonous materials should be kept in suitable packages with a label showing the name of the material, the supplying factory, and the quantity contained within. The word "poisonous" should be written in Arabic, as well as in one foreign language, in red and in a clear place on the packaging; and

5. The owner of the shop or store handling poisonous or non-poisonous chemicals used in an industrial environment must keep a logbook with its pages serially numbered and stamped by the ICA. All sales or other supply of such chemicals should be indicated in this book.

Decree No. 342/1962, which also amended Decree No. 138/1958, expressly exempts from licensing non-poisonous material(s) which are imported or purchased by factories for use in manufacturing their products.[208]

Law No. 21/1958, specifically through Chapter 2, Articles 14–15, "Concerning Organization and Development of Industry," authorizes the Ministry of Industry to establish specifications for raw materials and industrial products. Such specifications must be strictly followed in the production of more than 150 commodities.[209] Relatedly, Law No. 21/1957, "Concerning the Egyptian Organization for Standardization and Quality," directed the MOI to issue specifications for specific types of chemicals and household commodities, such as: red lead oxide primer, matches, paint solvents, fuel, pigments, dyes, food additives, perfumes, soap detergents, clothes, and blankets.[210]

A rapid expansion of industrial areas, coupled with growing industrial infrastructure and concurrent use of chemical substances on an increasingly wide scale, gave rise to a host of environmental challenges. For this reason, 1994's Law No. 4, "The Protection of the Environment" (Law No. 4), the prime environmental and chemical regulation within Egypt, was promulgated. Law No. 4 was implemented by the Prime Minister's Decree No. 338 (1995), and was later amended by Law No. 9 (2009).

> Law 4 of 1994 and its executive regulations (1995) define the roles and responsibilities of EEAA in order to avoid a conflict with exist[ing] laws, which include regulation of air pollution, control of hazardous substances, management of hazardous waste and control of discharges to marine waters. Nonetheless, responsibility [for] existing laws and regulations remained in the traditional ministries.[211]

As one might infer, Law No. 4 primarily relates to – and regulates – environmental aspects such as air, water, and noise pollution. There are sections, however, which address, at least tangentially, hazardous waste and pesticides. Examining Law No. 4, we first see that Articles 5 through 9 are largely administrative in nature, concerning such items as the administrative aspects needed to create the Environmental Protection Fund (Fund) (Article 7), the objectives to be supported by the Fund (e.g. the "[t]ransfer of low-cost techniques that have been proved to have been successfully applied" and "Financing the manufacture of types of equipment, devices and stations treating environmental pollutants") (Article 8), and the establishment of "an incentives System that the EEAA and competent administrative bodies may offer to the authorities, establishments, individuals, and others who undertake works or projects that protect the environment."[212]

Article 10 is where the more fundamental aspects of Law No. 4 begin, with that Article setting out the authority for the evaluation of the establishments requesting required permits, and directing that the same be evaluated with respect to their environmental impact.[213] Article 11 delineates the establishments to whom Annex 10 applies (see Annex 2), while Article 12 lays out the criteria for applicants to complete.

Article 13 instructs the EEAA to essentially incorporate the evaluation(s) of any experts they deem to be appropriate when reviewing the "environmental impact of an establishment intended to be constructed and for which a permit is being requested," while Article 14 requires the competent administrative body to notify the applicants of the decision taken, the means of notification, and lays out the appeal process for adverse decisions.[214] This appeal process is further discussed in Articles 15 and 16. Article 17 mandates the establishment and maintenance of a "register to record the extent of their [the successful applicant] establishment's impact on the environment," with the specific format noted, while Article 18 lays the responsibility for verifying conformity squarely at the door of the EEAA, with objective evidence obtained, on an annual basis, and provides remedies for managing non-compliant establishments.[215]

The remaining Articles of Law No. 4 are more specific in their nature, with respect to the types of items they manage or regulate. Articles 19 through 22 discuss the establishment of Environmental Monitoring Networks and Emergency Plans, as well as their management and implementation, while Articles 23 and 24 address the prohibition on the "hunting, killing, or catching of birds and wild animals, as prescribed in Annex 4 of these Executive Regulations," as well as possessing, selling, or intending to sell them.[216]

As with the "break point" of Article 10 above, Article 25, which begins Chapter 2 of Law No. 4, serves as another critical point in which specific requirements are established, this time with respect to hazardous substances and wastes. Article 25 notes that "[h]andling of hazardous substances and waste shall be prohibited unless a permit has been issued by the competent body according to the type and use of hazardous substance and waste as follows."[217] Of note is that while Article 25 explicitly requires a permit for the legal handling of hazardous substances and wastes, the ensuing Article 30 clearly states that "[i]t is strictly prohibited to import hazardous waste or to allow its transit through the territory of the Arab Republic of Egypt."[218]

Article 25 goes on to delineate the competent bodies and the types of wastes for which they have responsibility:[219]

1. Hazardous agricultural substances and waste, among which are pesticides and fertilizers – MALR;
2. Hazardous industrial substances and waste – MOI;
3. Hazardous pharmaceutical and laboratory substances and waste and domestic insecticides – Ministry of Health;
4. Hazardous petroleum substances and waste – Ministry of Petroleum;
5. Hazardous substances and waste from which ionized radiation is emitted – Ministry of Electricity; Authority for Nuclear Energy;
6. Hazardous, inflammable, and explosive substances – Ministry of the Interior;
7. Other hazardous substances and waste – the competent Minister for Environmental Affairs shall issue a decree determining the competent body for issuing such a permit upon the proposal of the EEAA Executive Head.

Readers should pay careful attention to item number 7, and recognize that hazardous substances and wastes which are not identified in the previous six points may still be subject to regulation and attendant permitting by decree of the Minister of Environmental Affairs. Each of the Ministries identified above must promulgate a table of such substances and waste, and identify the degree of danger of each, the mandatory control standards to be considered during handling, the means of disposing of empty containers which contained the substances, and any additional standards which they may decide are appropriate.[220]

Article 26 sets forth the "Procedures for Granting a Permit," including the roles and responsibilities of both the applicant and the Ministry/Ministries. "The individual or the establishment wishing to obtain a license for handling of hazardous substances or wastes shall submit an application containing the following data":[221]

- Information about the establishment and the handler of hazardous substances and wastes;
- [Information regarding the] producers of hazardous substances or wastes;
- A complete description of the hazardous substances or wastes intended to be handled;
- The amount of hazardous substances or wastes intended to be handled annually and a description of the method of packing;
- The means to be used in storing hazardous substances or wastes and the storage period;
- The available means of transport, their routings and schedules;
- A complete statement of the method intended to be used for the treatment and disposal of the hazardous substances or wastes for the displacement of which [sic] a license is sought;
- A commitment to keep registers containing detailed accounts of the sources, quantities, and types of hazardous substances or waste, the rates and periods of their collection and storage[,] and the means of their transport and treatment [as well as] to furnish such data on request and not to destroy the registers for a period of five years running from the date they are first opened;
- A commitment to take all procedures as are necessary to ensure the proper packing of hazardous substances or wastes during the collection, transportation, and storage phases and not to mix them with any other type of wastes, as well as placing [a] written description on the container and [having] a previous experience certification in this field; and
- A details [sic] description of the emergency plan for confronting all unforeseen circumstances which guarantees the protection of human beings and the environment.

The EEAA and/or the Ministry of Health may request the applicant to fulfil other conditions as it deems necessary to ensure the safe handling of these substances.

Such a permit is valid for a maximum of five years, at the discretion of the body issuing the permit. However, the licensing authority may revoke or suspend the license in the following cases:[222]

1. If the license was issued as a result of the submission of incorrect data;
2. If the licensee violates the conditions of the license;
3. If the performance of the activity results in dangerous environmental effects which were unforeseen at the time the license was issued;
4. The emergence of sophisticated technology, which may, with minor modifications, be applied, and the use of which would lead to a marked improvement in the environment and the health of the workers; and
5. If the EEAA concludes that handling any of the substances and wastes is unsafe.

Articles 27 through 33 relate to the following items regarding hazardous substances and wastes, respectively: permits, hazardous waste management, construction of hazardous waste treatment facilities, precautions to be taken at hazardous waste production or handling facilities, packaging specifications and attendant data required at the same.[223]

The remaining Articles of Law No. 4 encompass the specific requirements for, permissible levels of, and other items relating to the "Protection of the Air Environment from Pollution" (Articles 34 through 49), "Protection of the Water Environment from Pollution" (Articles 50 through 56), and "Pollution from Land Based Sources" (Articles 57 through 65).

Pesticide Regulations

> Egypt is primarily an agricultural country. Pesticides and fertilizers are being used extensively to increase [the] crop yield of limited cultivable land to meet the requirements of the exponential increase in population. The overuse of such substances contaminates the air, the soil, the surface and underground water, and the crops. Outdated pesticides and fertilizers and their empty containers create serious health and environmental problems.[224]

In a similar vein to the considerable history of chemicals management in Egypt, there were two key pieces of legislation which formally launched pesticides regulation in the country. These legislative enactments were preceded in 1957 by Ministerial Decree No. 10/1957, "Concerning Licensing of Household Insecticides," which addressed the topic.[225] In 1966, Agriculture Law No. 53 regulated the production, import, and use of pesticides and fertilizers. Key among the requirements of Law No. 53 were a definition of agricultural pesticides ("those chemicals and formulations used to control plant diseases, pest insects, rodents, weeds, [and] other organisms detrimental to plants, animal insects [sic] and parasites") and the creation of a Pesticide Committee, whose role was "to specify pesticides to be used in [the] country, determine their specifications, procedure of their registration [sic] and condition for use."[226] Further, Article 80 of Law No. 53 required the MOA to issue enabling regulations for agricultural issues, particularly:[227]

1. The kinds of pesticides to be imported for local use, their specifications, conditions of importation, and handling;
2. Conditions and procedures of licensing for pesticides [importation and sale];
3. Procedures of pesticides registration, registration renewal, [and] registration fees; and
4. Methods of pesticides sampling and analysis, ways of disapprobation [sic] by the producers on results of chemical analysis, procedures to be followed in considering approbation and judging its validity, and the fees to be paid for such approbation.

Finally, Article 82 of Law No. 53 made a notable linkage between the marketing and promotion of pesticides, and the product uses approved by the MOA. This is one of the first instances in the region in which pesticide product advertising was regulated by the MOA. Specifically, "[a]dvertising or distribution of information on pesticides should comply with its specification and conditions for handling and registration and also with the recommendations of the Ministry of Agriculture for their use."[228]

Decree No. 50 of the following year managed the toxic properties of pesticides and the procedures for documenting them.

The MALR holds responsibility for formulating policies with respect to the use of fertilizers and pesticides. The Agricultural Pesticides Committee (APC) of the MALR, constituted in 2001 by Ministerial Decree No. 2188, maintains a website (http://www .apc.gov.eg/EN/) which "contains all relevant information with regard to pesticides (Database for all pesticides recommended for use in Egypt – Instructions for the safe use of pesticides in Egypt – Database for pesticide traders in Egypt – Information about APC…)."[229]

The chief regulation concerning agricultural pesticide regulation and management in Egypt is Ministerial Decree No. 1018 (2013), "Concerning Registration, Handling, and Use of Agricultural Pesticides in Egypt" (Decree No. 1018). Decree No. 1018 rescinded Ministerial Decree No. 1835 (2011) on the topic.

Examining several key Articles of Decree No. 1018, we begin with Article 1, which defines the APC as "the sole statutory agency responsible for registration and handling of agricultural pesticides in the Arab Republic of Egypt."[230] This point is echoed later in Article 8, which states that "[p]roducing, formulating, repacking, importing, trading, handling or using of agricultural pesticides in their crude or formulated forms is pro- hibited unless these pesticides are registered with APC according to the conditions and procedures stipulated in the present decree."[231]

In Article 3, the APC "endorses every action that contributes to the rational use of pesticides and the implementation of IPM [Integrated Pest Management] policies and strategies," thereby laying out one of the central tenets of both their mission and of their policy direction. In the ensuing Article, the APC continues with its "policy statement" of sorts, noting that it specifically recognizes and incorporates the FAO definition of the term "pesticide," as well as the concept of "safety," "wherein safety implies that a pesticide must be safe to human health, the environment and the benignity of agricultural crops and products."[232]

Article 5 of Decree No. 1018 relates directly to the substances which the APC registers, and addresses their correspondence to a variety of other international regulatory agencies, stating that the "APC registers active ingredients of agricultural pesticides in their 'crude' or 'formulated' forms, according to the registered pesticide database of the U.S. Environmental Protection Agency or the European Commission, or any other international agency accepted by the APC."[233] This is a particularly noteworthy Article, as it lays the groundwork for future sections of Decree No. 1018. Deconstructing it, we see that the APC registers "active ingredients" for agricultural pesticides, in their "crude" (read as: chemically neat) or "formulated" (read as: prepared in the specific final product) forms, according to the listing of the active(s) in the U.S. Environmental Protection Agency's (EPA) registered pesticide database, the European Commission, or any "other agency" which the APC recognizes. As the United States and the EU are the only two jurisdictions specifically mentioned in Decree No. 1018, it is recommended that the APC be contacted before considering active ingredients not presently recognized by either of these two entities.

Article 6 grants APC the right to set restrictions on the handling and use of the pesticides it registers, in part by specifying the permitted amounts of such pesticides that may be used, as well as by regulating the means of handling, trade, and application. Here, the Article again specifically references the FAO, citing their "Code of Conduct."[234]

Article 7 mirrors to some degree the U.S. EPA's Registration Review concept, under which registered pesticide active ingredients are reviewed approximately every 15 years, noting that they "regularly" (time period undefined) review the "situation" of registered pesticides "or those under the process of registration and take appropriate actions in the light of any new development pertaining to the safety of those pesticides to human health, the environment and the benignity of agricultural crops and products."[235]

Article 9 of Decree No. 1018 lays out the steps and data requirements for an agricultural pesticide registration application. These are:[236]

1. The applicant must submit to the APC's Rapporteur a completed application, the "primary file of technical information" and a receipt demonstrating payment for each crude or formulated pesticide for which registration is desired. The "primary file of technical information" is defined in Article 12 to consist of:

   > Documents, data, technical studies, authentic certificate attesting to the registration and use of the pesticide in its country of origin, a quality assurance certificate for the pesticide formulation from the manufacturing company, all the information pertaining to methods of analyzing the pesticide and the impurities that may co-exist with it, values of its physical and chemical properties, risk assessment and all necessary documents issued by the responsible authorities.[237]

2. The APC's Rapporteur will examine the application against the requirements of Article 5 of Decree No. 1018, to verify completeness. It should be noted that Article 9 expressly states "[t]he submission of the application is considered to be an implicit approval of the applicant to comply with all the rules, regulations and procedures governing the implementation of this decree's provisions," which may serve as the grounds to bring regulatory or other action(s) against the registrant for violating the attendant requirements.[238]

3. As per Article 10, the agricultural pesticide for which registration is being sought should not have a generic or common name or code number, nor should it be similar or identical to an already commercially registered pesticide.

4. Article 11 of Decree No. 1018 is somewhat unique to pesticide registration statutes in the Middle East, in that it specifically addresses Intellectual Property Rights (IPR). Upon acceptance by the APC of the primary file of technical information, for compounds within the IPR protection period, the applicant must submit samples of the Active Ingredient (AI) as well as of the major impurities which may co-exist within it. These samples must be from certified sources, for use as reference standards. For compounds which are beyond the IPR protection period, the Central Agricultural Pesticides Laboratory (CAPL) bears responsibility for obtaining the samples identified previously from approved sources.[239] The CAPL, one of the research laboratories belonging to the Agricultural Research Center, was established in the 1950s as a joint project between the Egyptian government and the FAO.

   > According to the law of agriculture number 55 in 1966 and the ministerial law number 622 in 2008, the CAPL controls the importations, manufactures and formulators of pesticides. The central pesticide lab also checks the validity of pesticides according to specifications of the FAO and the WHO.[240]

5. The applicant must also submit samples of the pesticide for which the application is being made, in quantities determined by the APC. Once received, the APC will select the agricultural research stations which will conduct the studies (called "experiments" in the Decree) on the pesticides, often for two consecutive agricultural seasons (one for attractants), but this period may be modified.

6. If the pesticide successfully passes all the "experiments," then the APC will issue a "Biological Assessment Certificate," signed by the APC's Rapporteur, as well as the APC's Chairman or his designee. The certificate will be sent to the applicant, and a copy will be retained by the APC. Of note is that Article 21 empowers the APC to approve "off-label" use of one pesticide to control (one of) the same pest on a different crop.[241]

As per the direction of Article 15, and also as noted previously, pesticide registration certificates are issued by the APC for the crude product or the formulated pesticide. The APC may issue such certificates for one year. The certificates are valid for six renewable years from the date of issuance, but efficacy must be reassessed at the beginning of the third year of renewal.[242] The APC may approve the import of a pesticide which it has not registered for use on green areas or on farms producing "export-oriented" crops, provided that the pesticide is registered "in the database of any internationally recognized reference agency referred to in article (5) of this decree."[243] The "technical import permit" is issued under the seven conditions listed in Article 24. Similarly, Article 25 allows the APC to issue a temporary registration certificate and label for a pesticide not registered in Egypt, provided it is in the aforementioned database(s).

Registration certificates are only transferrable in the case of a corporate entity ownership transfer. In the event that the APC decides to suspend or cancel a registration, the "concerned person will be given a grace period until the end of the agricultural season that follows, without exceeding the expiration date of the registration certificate." Registrations may be suspended or cancelled when any of the following occur:[244]

1. Not fulfilling any of the registration conditions stipulated in this decree;
2. Reduced effectiveness of the pesticide against the target pest;
3. Occurrence of clear natural imbalance in favor of the pest;
4. Occurrence of unexpected harms of the pesticide on human health, the environment, or on the benignity of agricultural crops and products; and
5. Publishing scientific reports by well-trusted reference agencies indicating that the pesticide is hazardous to human health and the environment or exceeds its Maximum Residue Level (MRL) values on agricultural products.

The registration holder may appeal within 30 days of receiving the notification from the APC. The APC will resolve the appeal within 60 days, and its decision is final.

Labeling of duly registered agricultural pesticides must contain all the technical and instructive information, with its color following the toxicity classification of pesticides as recommended by the World Health Organization (WHO), and the label must be affixed to the pesticide package.[245] Agricultural pesticide labels are valid for two years, but the label validity cannot exceed the expiration date of the pesticide's registration certificate. Of note is the language of Article 17, which allows "the concerned person to correct errors without violating the essence of the procedures and regulations of registration and handling," thereby precluding a potentially onerous review process to amend the label.[246] Further, the concerned person is prohibited from publishing any information regarding the pesticide(s) "in the form of advertisement, poster, pamphlet or newsletter or any other means of advertising unless approved by the APC's Rapporteur," as per Article 45.[247] It is the responsibility of the publishing agencies to verify the APC's approval of the information prior to publication.[248]

Article 27 mandates that synthesis and repackaging of these pesticide products must be performed at establishments licensed under the decree, similar in concept to the U.S. Federal Insecticide, Fungicide, and Rodenticide Act's (FIFRA) "Registration of Pesticide and Active Ingredient Producing Establishments" (cf. 40 C.F.R. § 167.20 *et seq.*). The establishments are subject to evaluation by the CAPL, and the license will be valid for a renewable four-year term.[249]

Article 40 of Law No. 4 addresses the "[s]praying or using pesticides or any other chemical compounds for purposes of agriculture, public health or others shall be prohibited except after complying with the conditions, norms and guarantees set by the Ministry of Agriculture, the Ministry of Health, and the EEAA, as follows":[250]

A. It is mandatory to notify health units as well as veterinary units of the types of sprays and antidotes before pesticide spraying;
B. Necessary first aid supplies shall be provided;
C. Protective clothing and materials shall be provided for laborers carrying out the spraying;
D. Citizens shall be warned about being in sprayed areas;
E. Spraying shall be carried out by laborers trained for this type of work; and
F. Special consideration shall be given to refrain from spraying by air planes except in cases of extreme necessity as estimated by the Minister of Agriculture. In such a case, the areas that require spraying shall be indicated and determined on maps, and shall be highlighted with a special color along with marking the principal flying obstacles and the regions where it is prohibited to spray. These areas include the vicinity of residential areas, apiaries, fish farms, and poultry farms, as well as cattle sheds to guarantee that man, animals, plants, water courses, or other components of their environment shall not be exposed, directly or indirectly, currently or in the future, to the harmful effects of these pesticides or chemical compounds.

Occupational Safety and Health Regulations

"In Egypt, the safety and health of workers has been a legal matter of concern since the beginning of the last century. The earliest legislation pertaining to occupational health in Egypt dates back to July [of] 1909."[251] In 1959, the first comprehensive law on the topic, Act No. 91, also known as the Labor Law, was promulgated. Act No. 91 replaced a patchwork of labor regulations aimed at specific groups, and brought them under a central umbrella. Among these patchwork regulations were:

> Act No. 48 (1933) governing the employment of juvenile workers of both sexes in industry; Act No. 80 (1933) concerning the employment of women in industry; Act No. 147 (1935) fixing the number of hours of work in certain industries; Act No. 317 (1952) on individual contracts of employment; Act No. 46 (1958) organizing work in mines and quarries; and Act No. 14 (1959) governing vocational rehabilitation and employment of disabled persons.[252]

The management of OSH in Egypt falls under the jurisdiction of the MOMI, which implements the regulations through its Central Authority for Protecting the Labor Force and the Working Environment (CAPLFWE). CAPLFWE is divided into three General Administrations:[253]

1. The General Administration for Labor Inspection, dealing with the enforcement of the labor legislation related to employment conditions in general;
2. The General Administration for Occupational Safety and Health and the Protection of the Working Environment (GAOSHWE; responsible for drafting safety and health policy and programs, procedures, guidelines, and legal requirements); and
3. The General Administration for Manpower Services.

Furthermore, the MOHP maintains under its purview a Chemical Safety Unit, which holds OSH responsibilities for the following:[254]

1. Provision of tools and mechanisms for safe handling of chemicals throughout the following steps: importation, transportation, storage, use, and waste management;
2. Maintaining a national chemical registry for all used chemicals (imported or locally manufactured). *Note:* this is simply a registry of chemicals used in Egypt, not a national inventory requiring pre-registration or notification;
3. Evaluating chemical substances, utilizing preventive measures, and reporting to authorized agencies how citizens may avoid exposure to these hazards;
4. Establishing of six poison management and information centers in six governorates, and providing the centers with suitable equipment and laboratory facilities;
5. Implementation of a toxico-vigilance program for the different levels of health care facilities in the different governorates;
6. Supplying the Unit with audiovisual materials on chemical safety for professional training; and
7. Participation in the Eastern Mediterranean Regional Office (EMRO) of the WHO plan for preparedness to chemical accidents.

Presently, Law No. 12 (2003), issued and published in the Official Gazette No. 14 on April 7, 2003, is the governing law organizing employment relations, clarifying the duties and rights of the parties to the employment agreement, and ensuring OSH in the workplace.[255] Somewhat uniquely, Law No. 12 mandates the establishment of an "OSH Committee" in workplaces of greater than 50 employees, with composition and function specifically defined in Decree No. 134 (2003), which will be discussed later. Key among these is that the Committee shall study the working conditions and causes of accidents and diseases, and shall specify preventive measures to reduce or eliminate them. The employer then has the responsibility to implement these recommendations. In situations where there are separate workplaces belonging to the same establishment, a central OSH Committee must be established at the headquarters location. It should be noted, however, that Decree No. 134 does not require employers to make a risk assessment of the work processes, machines, and equipment used by the employees as part of the work operations.[256]

Perhaps interestingly, violations of OSH regulations in Egypt tend to incur heavier penalties compared with violations of other Egyptian regulations.[257] Such penalties are set out in Book VI of Law No. 12 (2003). The two features of Egyptian OSH penalties are that they may include imprisonment and/or a higher minimum level of pecuniary fine.[258]

Imprisonment may be levied for a term of not less than three months and a fine of not less than 1000 Egyptian pounds (LE) for infringement of Articles 202–231, while a fine of not less than 500 LE and not exceeding 10,000 LE applies for infringement of Articles 234 and 235.[259] Additionally, penalties may be doubled in situations where the violation is a repeated one.[260]

Chemical OSH regulations are partly managed by Law No. 12, as well as by Decree No. 211 (2003) and Law No. 4 (1994), the last of which has been discussed previously. Law No. 12 entered into force on July 7, 2003, replacing Law No. 137 (1981) and Ministerial Decrees Nos. 33/1982, 119/1982, 105/1987, 142/1988, 145/1988, and 10/1991.[261] Law No. 12 is supplemented by Ministerial Decrees which define the more specific technical provisions, such as:[262]

1. Decree No. 126 (2003), replacing Decree No. 75 (1993), defining procedures and forms for the notification of work-related accidents, injuries, fatalities, and diseases;
2. Decree No. 211 (2003), replacing Decree No. 55 (1983), specifying conditions and precautions essential for the provision of OSH measures at the workplace; and
3. Decree No. 134 (2003), replacing Decree No. 116 (1991), defining the types of establishments covered, OSH services and committees, and related OSH training institutions.

Law No. 12 applies to all establishments in the private and public sectors, civilian government units, local (municipal) government services, and public authorities, specifically under Article 203.[263] Perhaps interestingly, it does not apply to household servants and family members who are direct dependents of the employer.[264] Book V of Law No. 12:

> applies to all branches of industry, including the construction industry, commercial establishments, and agriculture. It applies to all working sites and establishments, once authorized, whatever the number of workers employed. Specific provisions apply to establishments with more than 15 to [less than] 50 workers.[265]

Some key Articles of Law No. 12 (2003) are as follows:[266]

- Article 208: Employers must take all needed measures to ensure safety and health in the workplace, specifically regarding mechanical, physical, chemical, and biological hazards;
- Article 216: A medical evaluation of the worker must be conducted before employment. Medical attention and treatment must be provided during the course of work, although this aspect depends upon the number of workers employed. Periodic medical examination of workers who are exposed to the risk of any occupational diseases as set out in Annex I (2004) of Law No. 79 shall be undertaken; and
- Articles 208–215: It is incumbent upon employers to inform employees of the dangers they are exposed to, and to provide workers with Personal Protective Equipment (PPE).

The MOMI, as well as the local Council authorities responsible for manpower, have the sole authority to inspect establishments for OSH compliance.[267]

In addition to Law No. 12 (2003), the protection of workers against hazardous processes, machinery and equipment, hazardous chemicals, and physical and biological agents is regulated by three major Decrees:[268]

- Decree No. 126 (2003) sets out the procedures and forms for notification of accidents and diseases in the workplace. It also specifies the type of statistics which should be collected and reported to the Authority with regard to major injuries and accidents;
- Decree No. 211, also of 2003, lays out the conditions required for a safe working environment, with respect to physical, mechanical, electrical, chemical, biological, and other hazards. Various chapters within Decree No. 211 (2003) establish the "Maximum Allowable Concentrations" for "more than 600 chemical agents in the working environment, safe levels of physical parameters (heat/cold, stress, noise, vibration, illumination, radiation, static electrical fields, classification of jobs according to physical workload, etc.), and a list of suspected chemical carcinogens"; and[269]
- Decree No. 134 (2003), as noted above, defines the type of enterprises which should have an OSH Committee. "It also regulates training in occupational safety and health for workers/managers involved with OSH in the enterprise… The main functions of OSH technicians and specialists are":[270]
  1. Periodic inspection of the workplace;
  2. To investigate accidents and determine their causative factors;
  3. To investigate the incidence of occupational diseases and determine their causative factors;
  4. To maintain statistical information;
  5. To check fire fighting [sic] equipment and follow up protective measures;
  6. To participate in safety committee meetings; [and]
  7. To specify preventive measures.[271]

As per Decree No. 134, the OSH Committee should consist of the following representatives:[272]

1. The facility owner, his representative, or the General Director;
2. The heads of the main production sections or departments;
3. A representative of the Civil Defense [Authority];
4. The facility physician, if one exists;
5. The person in charge of OSH at the facility; and
6. A number equal to [the] above members [inferred to mean equal to the number of OSH Committee members formed from items #1–#5 above], from local trade union members, and selected from the same production sections and/or departments, also as above.

Three other Decrees are worthy of note with respect to OSH management in Egypt. Decree No. 55 of 1983 regulates all chemicals used in the industrial field, specifically the control, use, handling, and storage of chemicals and conditions required for industrial safety and health in the workplace.[273] Decree No. 116 (1991) strengthens the facilities with training for directors and workers, and Decree No. 499 (1995) regulates the handling of poisonous and non-poisonous substances in industry.[274]

Waste Regulations

The management of hazardous waste has been identified by the government of Egypt as a priority issue since at least 2004. "Hazardous waste" is defined in Law No. 4 (1994) as:

Waste of activities and processes or its ashes which retain the properties of hazardous substances and have no subsequent original or alternative uses, like clinical waste from medical treatments or the waste resulting from the manufacture of any pharmaceutical products, drugs, organic solvents, printing fluid, dyes and painting materials.[275]

Law No. 4 was the first piece of Egyptian legislation to expressly acknowledge hazardous waste as a "potential source of significant environmental degradation."[276] Although Law No. 4 clearly defines the term "hazardous waste," neither Law No. 4 nor the enabling regulation, Prime Minister's Decree No. 338, specify the objective criteria which makes such a waste hazardous (e.g. pH, corrosivity, flash point, constituent substances, etc.), nor does it identify specific waste streams which may be hazardous by their nature (cf. the United States' RCRA "F" and "K" listed wastes at 40 C.F.R. § 261.31 and 40 C.F.R. § 261.32, respectively). Rather, Law No. 4 "specifically address[es] different components of HW management through a number of stipulations concerned with waste minimization, waste storage, packaging, collection, transportation, treatment, and final disposal."[277] Such ambiguity was the result of a "compromise" among the various Ministries involved in implementation of Law No. 4 (see above) – each would have the ability to specify the type(s) of waste it considers to be hazardous within its jurisdiction, even in the absence of an overarching classification framework.[278]

As time progressed, the challenges inherent in the foregoing became more apparent, causing the EEAA to develop a national system for hazardous waste identification, as well as a national system of "best practices" management for the same. Thus, the national system for the identification and classification of hazardous waste was developed, which was based on the U.S. EPA's RCRA hazardous waste definitions and classifications, as set forth in 40 C.F.R. § 261.1 *et seq.* In short, a waste is defined to be hazardous under RCRA if it is either a "listed waste" (specifically listed in the regulation as such), or is a "characteristic waste," meaning it has one of four specific characteristics: ignitability, corrosivity, flammability, or toxicity (a concentration of a specific substance above a specified amount). Subsequently, "[o]perational guidelines were developed for different components of HW [hazardous waste] management... these guidelines offer best practices used as standards for the assessment of HW management activities... They have been developed in light of practices and experiences available internationally, but adapted to the Egyptian conditions."[279]

Even as the means of identifying and classifying hazardous waste in Egypt was developing, a new hurdle in its management emerged – how to properly dispose of wastes identified as hazardous. Indeed, "the substantial financial requirements needed to establish central HW treatment and disposal facilities, together with central budgetary limitations, on the other hand, significantly delayed the establishment of such facilities through public funds."[280] Complicating the funding of such facilities was the fact that the scale of waste to be managed was simply not known. Thus, the planning for facilities to dispose of the waste was speculative at best.

Furthermore, while Egypt now had a means to identify and classify hazardous waste produced in the country, the lack of established, well-defined, and efficient hazardous waste treatment and disposal facilities put companies attempting to comply with the regulations into a challenging situation:

Generators realized that if they identified what waste they generated, segregated it from non-hazardous streams, properly stored it on site, and for some waste, pre-treated it on site, they would be unable to dispose of it in any legal way! They would become liable for waste which they could not get rid of![281]

As a consequence, most hazardous waste generators evaded identifying their waste, with the thought process being that failing to have segregated and defined hazardous waste on-site would make their compliance status less clear.[282]

Similarly, the Ministries – who were responsible for enforcing the hazardous waste regulations, including disposal – could not enforce the regulations due to the lack of (compliant) treatment and disposal facilities. However, and perhaps ironically, there was then no impetus to create such facilities, since there was no enforcement![283]

Given this untenable environment, the Egyptian authorities began a waste minimization program, focusing first on the industrial sector. By coordinating these activities with a variety of international funding instruments and technical assistance, the program began to develop momentum within the regulated community.[284] It, however, relied on the desire of hazardous waste generators to take the initiative to participate, and to implement the recommended minimization efforts. "In order to successfully achieve these measures, such establishments have to identify their HW, segregate it from non-hazardous waste, and where applicable further implement other waste management measures."[285]

At present, the Egyptian authority has seen a marked trend toward hazardous waste minimization in the country; however, as the industrial environment continues to grow, the question remains as to whether this voluntary compliance, which comes at an administrative cost to the companies involved, will continue to take precedence over non-compliance, which currently results in little to no enforcement.

Safety Data Sheets and Labels

Egypt terms its SDSs "Emergency Response Sheets" (ERS). There is no formal requirement for the ERS, although the application process under EHSIMS directed that they should be submitted along with other required documentation.[286] ERS include, among other items, the chemical name and classification of the substance, the attendant CAS and UN (dangerous goods) number(s), health hazard potential and Personal Protective Equipment (PPE), first aid statements, disposal information, and emergency telephone numbers.[287] At present, the Globally Harmonized System of Classification and Labeling (GHS) has not been implemented in Egypt, and there is no implementation foreseen.[288]

Hazardous chemical substance labeling in Egypt contains multiple elements which must be followed to ensure compliance, as laid out in Article 32 of Law No. 4. Specifically, these are:

1. Container specifications:
   The type of container in which these substances are placed must be suitable for the type of substances therein, tightly closed, difficult to damage, easy to lift or transportation without exposing it to damage or harm.[289]

2. Container information:

   Content of container, their active substance, the degree of its concentration, total and net weight, and name of producer, date of production, production number, nature of danger, symptoms of toxicity, first aid procedures, safe storage methods, and methods of disposal of empty containers.[290]

   All information provided on the labels must be written in Arabic in a style that is easy for an ordinary person to read and understand. Furthermore, the words must be legible and prominently displayed on the container.[291] The labels shall be accompanied by diagrams which provide instructions regarding how to open, empty, store, and dispose of the containers, as well as including the UN Dangerous Goods pictograms for danger and toxicity.[292]

4

Israel

National Overview

The State of Israel (Israel) was created on May 14, 1948 by the UN. "After the Nazi Holo-
caust, pressure grew for the international recognition of a Jewish state, and in 1948 Israel
declared its independence following a UN vote to partition Palestine."[293] "On November 29, 1947, the United Nations adopted Resolution 181 (also known as the Partition
Resolution) that would divide Great Britain's former Palestinian mandate into Jewish
and Arab states in May 1948 when the British mandate was scheduled to end."[294] Israel,
which began as an agrarian collectivist state, has undertaken a rapid journey to that of
a high-tech economy over the previous 70 years.[295]

The country consists of six administrative districts (*mehozot*): Central, Haifa,
Jerusalem, Northern, Southern, and Tel Aviv.[296] Israel has four major political parties: Kadima, Likud, Yisrael Beytenu, and Labor. There are also eight other parties
represented to varying degrees in the legislature.[297]

Before moving further, it should be noted that most regulations in Israel have two dates
as part of their nomenclature; the first being the Jewish calendar year, and the second
being the conventional Gregorian calendar. For example, the Hazardous Substances Law
is codified as "5753-1993."

Governmental Structure

The governmental structure of Israel is that of a parliamentary democracy, consisting
of executive, legislative, and judicial branches.[298] Israel currently has no formal,
single-document constitution; however, some functions of a constitution are filled by
the Declaration of Establishment (1948), the Basic Laws, and the Law of Return (as
amended).[299]

The Executive branch is headed by the President, with the Prime Minister as the
leader of the government.[300] The President is elected to a seven-year term in office,
but this is primarily a ceremonial role. "The Prime Minister functions as the head of
Government and exercises executive power."[301] He/she is elected by the Knesset, the
unicameral national legislature of Israel. The Prime Minister must be a member of the
Knesset, and requires only a simple majority of votes to be confirmed. Prime Ministers
serve four-year terms. To form a new government, a prospective Prime Minister has
45 days to fill Cabinet positions, and to win Knesset approval of the nascent government.

Chemical Regulation in the Middle East, First Edition. Michael S. Wenk.
© 2018 John Wiley & Sons Ltd. Published 2018 by John Wiley & Sons Ltd.

The Legislative branch is the Knesset (Parliament), which consists of a single, 120-member chamber.[302] Members of Parliament are directly elected by the people to four-year terms. Additionally, "seats in… the Knesset… are assigned through a system of nationwide proportional representation: Rather than electing individual candidates, voters cast ballots for an entire party. Any party receiving more than 2 percent of the vote is assigned a proportional number of seats…"[303] "Legislation may be presented by an individual Parliament member, a group of Parliament members, the Government as a whole, or a single Minister from within the Government."[304]

Israel's legal system is based on common law, "the body of law developed in England primarily from judicial decisions based on custom and precedent, unwritten in statute or code, and constituting the basis of the English legal system."[305] "The prevailing characteristic of the legal system, however, is the large corpus of independent statutory and case law that has evolved since 1948."[306] Thus, Israeli law is a mixed legal system of English common law, British Mandate regulations, Jewish, Christian, and Muslim religious laws.[307]

Judicial branch power is vested in the Judiciary, which is divided into the general courts of law, also called the civil courts. The civil court system is comprised of three levels: the Supreme Court, the District Courts, and the Magistrates' Courts. The Supreme Court, the highest Israeli court, consists of the Chief Justice and 14 judges.[308] Judges may serve up to mandatory retirement at age 70.[309]

Additionally, there are tribunals or other authorities of limited jurisdiction. These tribunals include the military courts, the labor courts, the administrative courts, and the religious courts.[310] The jury system is not employed in Israel.

Most primary laws are available in English, via authorized translations by the Ministry of Justice. The Official Journal (*Rashumot*) is the main source for all legislative and administrative actions in Israel, and is available at http://www.justice.gov.il/En/Pages/default.aspx.

Key Chemical Regulatory Agencies

Since 1993, the Ministry of Environmental Protection (Ministry) has been the Authority responsible for hazardous substances regulations in the country.[311] This management includes the classification of such substances in accordance with use, toxicity, and/or risk, as well as the manufacture, import/export, packaging, commerce, issue, transfer, storage, maintenance, and use.[312] At the local level, 36 "municipal environmental units" and "associations of towns for environmental quality" operate throughout the country. "These environmental units operate under the administrative jurisdiction of their respective municipalities but under the professional authority of the Ministry of Environmental Protection (MEP). They play an essential part in the provision of environmental services on the local level."[313]

Over time, however, other government agencies and Ministries have gained increased authority and power around environmental regulation. Among these are:

1. The Ministry of Industry, Trade and Labor and the Ministry of Agriculture and Rural Development. Both Ministries are becoming increasingly involved in regulating chemicals;

2. The Chemical and Environment Administration (CEA) of the Ministry of Industry, Trade and Labor. The CEA has an interesting dichotomy of purpose – it is responsible for industrial waste management as well as representing industry in any regulatory forum;
3. The Ministry of Agriculture's Plant Protection and Inspection Service. This division of the Ministry has responsibility for the registration and regulation of pesticides for plant protection;
4. The Ministry of Agriculture. This Ministry oversees and regulates quality and health requirements of exported agricultural produce. It also cooperates with international bodies relating to the standardization of pesticide tolerance regulations; and
5. The Ministry of Finance. This Ministry is responsible for the allocation of additional funds for the implementation of regulatory policy.

Key Chemical Substance Regulations

Israel's chemical industry is rooted largely on those fields in which it has a relative advantage, such as products based on minerals found in the country (e.g. magnesium, bromide, phosphates, potassium), a petrochemical industry, which is based on oil refining, and a pesticide industry. In addition, Israel produces organic intermediates based on local research and development.[314] The most significant segment of the Israeli chemical industry is the pharmaceutical business, which is a world leader in the manufacture of generic drugs.[315] "The main chemical industries are concentrated in three areas: Haifa Akko, in the north of the country, Ashdod in the center and Beersheba, Dead Sea, Mishor Rotem and Ramat Hovav in the south."[316]

Chemical regulation in Israel is fairly well advanced. "Enforcement includes supervision on the sales and acquisition of chemicals and supervision on the import of chemicals (by Israeli customs). Moreover, in recent years, an Integrated Pollution Prevention and Control approach was introduced into the major industrial hotspots: Ramat Hovav (*Ne'ot Hovav*) in the south, Ashdod on the southern Mediterranean coast and Haifa Bay in the north of the country."[317]

> Currently, registration and licensing of chemicals is carried out on pesticides, pharmaceuticals, cosmetic preparations and food additives. There is no legal framework for the licensing of industrial chemicals (e.g. a national inventory of existing or new chemicals), nor is there presently a new chemical notification scheme, although a licensing framework does exist and operates for chemicals used in agriculture (pesticides and biocides).[318]

Israel has adopted an extremely robust chemical regulatory structure similar to that in use in many of the Western countries. For example, Israel has codified a significant number of "Safety at Work" regulations, such as standards for isocyanates, benzene, vinyl chloride, arsenic, mercury, lead, and others.

Risk regulation, with special focus on environmental aspects, is the leading regulatory approach to chemicals management in Israel.[319] "A risk management manual which relates to risk reduction from industrial sources involving chemicals and to minimizing their impacts on public health and the environment was published by the MEP."[320] As per the manual, any premise which is subject to a Hazardous Materials Permit for a regulated substance and is included in a list of especially toxic or flammable substances is required to prepare a risk management plan. The risk management plan is largely based on the United States' State of California Accidental Release model (CalARP), which itself was implemented on January 1, 1997.[321]

Israel's environmental legislation, defined as encompassing specific laws regarding the pollution of air, water, marine, and noise, as well as waste, is wide-ranging: it seeks to prevent environmental deterioration while simultaneously stopping, abating, and remediating existing pollution.[322] National legislation is complemented by a wide range of environmental by-laws on the local level, and by an increasing number of international conventions on the global scale.[323]

Currently, the existing frameworks for industrial chemical management tend to focus primarily on the "supervision of the production, import, storage, processes, wastes and transport of chemicals."[324] However:

> There are several laws and dozens of regulations – created under the authority of those laws – that relate to public health and safety. These include legislations that relate to general environmental protection, public health, asbestos, hazardous materials, work safety, and radiation.[325]

As will be seen, chemical substance regulation in Israel falls primarily under two comprehensive laws: the Licensing of Businesses Law of 1968 and the Hazardous Substance Law of 1993. The regulation of pharmaceuticals and cosmetics chemical preparations, however, is regulated under the Pharmaceutical Ordinance of 1981.[326]

The State of Israel has signed and/or ratified several international treaties on the use of chemical substances:

a. The Stockholm Convention on Persistent Organic Pollutants (POPs): the pesticides listed in Annex A and B of the convention are banned in Israel;
b. The Rotterdam Convention on Prior Informed Consent Procedure for Hazardous Chemicals and Pesticides (PIC);
c. The Basel Convention on the Transboundary Movement of Hazardous Waste;
d. The Vienna Convention on the Protection of the Ozone Layer and the Montreal Protocol on Materials that Deplete the Ozone Layer; and
e. The Strategic Approach to International Chemical Management (SAICM).[327]

The existing main frameworks for chemicals management in Israel are managed through the licensing of businesses and hazardous materials permits, the latter of which is enshrined in Section 3(a) of the "Hazardous Substances Law" (HSL), the main legislation relating to the management of dangerous chemicals. The HSL was promulgated in 1993, and amended in 1997, 2002, 2004, 2005, and 2008. The HSL provides the Ministry with the necessary authority to manage substances deemed "hazardous." Further, the HSL details the process by which permits will be issued.

Annex I of the HSL lists 29 "harmful" substances and mixtures, and Annex II lists 219 "toxic chemicals." Both Annexes regulate the user of these chemicals. They provide stringent measures for supervision, "from cradle to grave," of the production, import, storage, processes, wastes, and transport of chemicals.

Under the HSL, "no person shall deal with hazardous materials, unless he holds a hazardous materials permit from the Supervisor."[328] As an aside, in practice, the "hazardous materials permit" is referred to as a "toxins permit."[329] The HSL defines a "hazardous material" as "each of the substances specified in Schedule Two, whether in its simple form or mixed with or blended into other substances."[330] Note that this differs from a "harmful chemical," defined as "each of the substances specified in Schedule One, whether in its simple form or mixed with or blended into other substances."[331]

The HSL is extremely broad in its scope, as it permits the Minister to "make regulations on (1) the handling, use, production, import, export, packaging, commerce, issue, transfer, storage and possession of hazardous substances" as well as "any other matter under this Law, which requires regulation."[332] The HSL directs that the Minister may "classify hazardous materials by the purpose for which they are used, by the degree of their toxicity, [and] by the degree of danger involved in their use or according to other considerations," and must consult other Ministers as appropriate (e.g. the Minister of Agriculture for substances used in agriculture, the Minister of Industry and Trade for industrial substances, etc.).[333]

Further, as per the HSL, only persons holding a hazardous materials permit from the Minister may handle such chemicals, and the permit is only valid for a specified period of time in accordance with criteria such as the "type of enterprise, the type of hazardous material and the quantity of hazardous material."[334] In addition, every place which sells "hazardous substances" must obtain a license under the Licensing of Businesses Law 5728-1968 (Business Law), which will be discussed in more detail later in this section.[335]

The requirements for the license are laid out in Section 6B(a), (b) of the HSL; however, the Licensing Authority may require additional documents and/or documentation not prescribed in that section, but which, in their opinion, are necessary for the application.[336] The position of the Minister must be communicated within 45 days of receiving an application (from the delegated Minister, as appropriate). If the Minister does not communicate his position within this time, it will be concluded that the Minister has given consent.[337] During the review process, the applicant has an absolute right to request the Authority to provide a preliminary opinion as to whether the applicant meets the requirements of the HSL, as well as if there are (expected to be) any conditions to be met prior to approval.[338] Moreover, the Minister can place certain conditions on the granting of the permit, as well as setting a fee for the permit and/or its renewal.

Finally, Section 6(a) of the HSL mandates specific storage requirements for hazardous materials ("shall be stored under lock and key and shall be kept separately from substances that are not hazardous").[339]

> Permits specify the types, quantities and compositions of chemicals in the facility. They also impose restrictions on the quantities and conditions of use of these materials for the protection of man and the environment. The holder of a permit is required to maintain a hazardous substances register in which details of all sales and purchases of hazardous substances are recorded. These record books facilitate tracking of the movement of hazardous substances throughout the country and illegal trade. A fully-computerized system classifies all permit holders into sectors and categories.[340]

In a 2004 amendment to the HSL, the disposal of hazardous substances waste was incorporated. Section 14 sets out the definition of "disposal," as well as the criteria by which disposal fees will be established, and exemptions for certain categories of persons which may apply. Of interest is Section 14M, which specifically allows the Director to *require* that the violator publish notice of the violation in a newspaper or another media which the Director decides.

The HSL is complemented by multiple regulations, such as:

1. Safety at Work Law No. 5730-1970, as amended;
2. Hazardous Substances Regulations (Classification and Exemption) No. 5756-1996;
3. Hazardous Substances Regulations (Criteria for Determining Validity Period of Permits) No. 5763-2003;
4. Licensing of Business Regulations, 1993 (as amended); and
5. Israeli Standard IS 2302-1-2004 on the classification, packaging, labeling, and marking of dangerous materials.

Finally:

> The Information and Response Center for Hazardous Substances collects quantitative and qualitative information on hazardous substances that are used, produced, imported, exported, transported, and disposed of in Israel. Data relate to quantities, types, characteristics, and concentrations of substances found in all sectors and at all levels, including industry and institutions on the local, regional, and national levels. In addition, the Center maintains data on safety, detection, identification, treatment, and neutralization procedures for chemical accidents.[341]

It should be noted that no assessments are performed, and the focus of the Center is limited to hazardous substances, thus excluding industrial, pharmaceutical, and agricultural chemicals.[342]

The Licensing of Businesses Law 5728-1968 (Businesses Law) directs the Ministry of the Interior, in consultation with relevant Ministries, to issue business licenses. Through the issuance of such license(s), specific conditions may be imposed, such as those relating to specific environmental goals in the areas of air quality, solid waste, hazardous substances management, and water and sewage (including industrial effluents).[343]

Partly as a result of such relatively broad authority, the Ministry of the Interior has allowed for classification of all entities appearing in the "Business Licensing Order" into one of three categories – A, B, or C – according to their potential environmental risk. The A category is the most severe, designating industrial plants with the greatest potential for environmental pollution. Local environmental units and associations of towns have been granted authority under the Businesses Law to manage C category businesses, including preparation of conditions and enforcement of their implementation.[344] It should be noted that in situations where a business requiring a license is also required to hold a separate license under another law, "the business license may be withheld until licensing under the other legislation is completed. For example, businesses handling hazardous substances are required to obtain Poison Permits (under the Hazardous Substances Law) before they are granted a business license."[345]

With the groundwork for the key chemical substance regulation in Israel laid, attention can now be turned to how such regulation is applied in various other areas of the chemical industry. In 1993, the Hazardous [Industrial] Plants Regulation (Regulation) was promulgated, under the authority of Sections 9, 10, 11A, and 39 of the Licensing of Businesses Law. The Regulation required "owners" of "industrial plants" "in which hazardous substances are stored, sold, processed or produced to take all necessary measures to treat these materials according to the best available technology and manufacturer directions."[346]

To fully understand the obligations imposed by this phrase, several of its terms must first be defined. First, a "hazardous plant" is defined in Section 1 of the Regulation as "a business that requires a license… in which hazardous substances or waste from such substances are stored, sold, processed or produced, or in which hazardous substances are produced during its processing or production process."[347] Next, many of the aspects of the Regulation apply to the "owner of a hazardous plant," although the term may be somewhat misleading, as it does not necessarily refer to the individual(s) who have legal ownership of the facility. Rather, it is defined in Section 1 as:[348]

1. The holder of or applicant for the license, as the case may be;
2. The person under whose auspices, supervision, or management the business operates; and
3. The registered owner or occupant of the property, in which the business is located.

Thus, the individuals and groups which may have interactions with – and various types of responsibilities for – such facilities should carefully review Section 1 to ensure they understand the responsibilities which may pertain to them. Finally, the Regulation defines "hazardous substances" perhaps somewhat differently from other countries, such as the United States, where such substances are often specifically delineated in national law [cf. the RCRA at 40 C.F.R. § 261.31]. Here, the term is defined to include:

> [A] substance, in any state of aggregation, which has a U.N. number as specified in the Orange Book [the U.N. publication "Recommendations on the Transport of Dangerous Goods"] and as described in Parts One and Two of Schedule One to the Goods and Services Control Order (Transport Services and Use of Trailers) 5739-1978.[349]

The Regulation lays out the measures that must be taken to "prevent and/or treat accidents, such as leaks, dispersal or conflagration of hazardous substances."[350] Furthermore, only skilled and specially trained personnel are permitted to handle hazardous materials at these establishments. The definition of "skilled and specially trained personnel" is found in Section 2.2 of the Regulation, but it is extremely vague: "the owner of a hazardous plant shall only handle or allow the handling of hazardous substances in his plant area by skilled manpower, which underwent appropriate training."[351]

As per Section 4(a), owners of plants subject to the Regulation must prepare and maintain files which include data on "the handling of mishaps and incidents, which are liable to occur in the course of operating his plant, and which are liable to pose danger to persons and the environment."[352] Note that the definition includes the conjunction "and"; that is, the danger to persons and the danger to the environment are not mutually exclusive. To satisfy the language of the Regulation, the "mishaps and incidents" must cause danger to <u>both</u> entities.

Section 4(b) sets out the items which are to be included in such a plant file:[353]

1. A plan and description of the plant, including a detailed list of the hazardous substances, their marks [hazard markings], and the methods by which they are handled;
2. A detailed list and definitions of the mishaps and incidents liable to occur in the course of the operation of the plant;
3. Existing means in the production system against mishaps and incidents due to the explosion, conflagration, or dispersion in the environment of hazardous substances;
4. Safety measures within the plant, including means of alert, means of neutralization, personal equipment and protective gear, and a fire detection and extinguishing system; and
5. The plant's preparedness plan for handling mishaps and incidents said in paragraph (2), which shall include existing means of neutralization and their manner of operation, a detailed list of the skilled manpower for the said treatment, a work plan for the activation of the manpower and equipment system, as well as particulars on means of communication and reporting to the competent authorities.

Owners of subject plants must also submit a copy of the file to the Licensing Authority, as per Section 4(c), and copies of the file, which should be "updated from time to time," shall be provided. This file is different from the "report" specified in Section 5 of the Regulation, which should be submitted to the Licensing Authority once a year. Notably, the Regulation allows the plant owner to determine when the submission shall be made – "at the time it determines and at additional times, according to needs and on demand."[354] The report should "include complete and updated details about":[355]

1. The categories of hazardous substances, their quantities, and the manner in which they are used;
2. Particulars about changes in the method of production and in the manner of using hazardous substances in production processes;
3. The manner of keeping hazardous substances within the plant, including details on storage conditions (such as temperature and pressure), categories of packaging, means of separating different categories of materials, as well as particulars about the storage area and its maintenance, including means of access;
4. Existing means in the production system to protect against mishaps and incidents in consequence of the explosion, conflagration, or dispersion in the environment of hazardous substances;
5. Safety measures within the plant, including means of alert, means of neutralization, personal equipment and protective gear, and the fire detection and extinguishing system;
6. A description of the plant's emissions to the environment, including the composition and quantity of sewage; and
7. Mishaps and incidents that occurred during the reporting period.

As per Section 6(a) of the Regulation, "the particulars included in the plant file said in regulation [section] 4" may be certified via an affidavit from the owner of the plant. The Licensing Authority, however, as per Section 6(b), has responsibility to provide copies of the report(s) and of the plant file to the "relevant factors under the law [regulatory agencies], at their request, to the extent relevant to their tasks and spheres of responsibility."[356] A violation of any provision of the Regulation shall incur six months' imprisonment, or a fine as per Section 61(a)(1) of the Penal Law 5737-1977.

Another regulation which arose from the HSL, specifically as per Sections 3(d), 10(1), and 12, is the "Hazardous Substances Regulations (Criteria for Determining Validity Period of Permits), 5763-2003" (Criteria). Recall that under the HSL, "Poisons Permits" are required for facilities that deal with hazardous substances. The manifest purpose of the Criteria was to promote the efficiency and strengthen the implementation and enforcement aspects of the permitting system.[357]

As the name suggests, "the purpose of these regulations is to determine criteria for the validity period of permits," with "permit" having been defined in the HSL.[358] Section 3(a) defines the "validity period of a permit to deal with hazardous materials" as being set out in Table B of the Schedule, which concludes the Permits regulation. Tables A and B of the Permits regulation relate to the "quantities, business types and activity types" for which a permit has been granted.

The classification into levels is directed in a "triage-type" manner in Section 3(b). To classify, the "level shall be initially determined according to Table A of the Schedule."[359] If Table A is found to not provide "an appropriate classification," then "the level shall be determined according to Table B of the Schedule."[360] In the case where a particular occupation may fit into more than one level in each table, Section 3(b)(3) directs that "the stricter level among them shall be determined."[361] That is, the stricter level of the two should be the one selected.

There are four lengths of permit validity as specified in Section 3(a):[362]

- Level A: one year;
- Level B: two years;
- Level C: three or more years; and
- "Under the Appointed Supervisor's decision – for a one-time occupation": For a period of less than one year.

Examining Table A of the Schedule, and using anhydrous ammonia as an example, we can see how the authority categorizes the substance into Level A, Level B, or Level C:[363]

Schedule (Regulation 3)

Table A Classification by types of hazardous materials, business and activity types, and quantity.

Column A	Column B		
Types of hazardous materials	Level A	Level B	Level C
Anhydrous ammonia	In a quantity of over 5 tons	In a quantity of 1–5 tons	In a quantity of under 1 ton

Source: Hazardous Substances Regulations, http://www.sviva.gov.il/English/Legislation/Documents/Hazardous%20Substances%20Laws%20and%20Regulations/HazardousSubstancesRegulations-CriteriaForDeterminingPermitValidityPeriod-2003.pdf.

Table B Classification by hazardous substances class.

Column A	Column B		
Nature of activity and hazard classes	Level A	Level B	Level C
Combustible Gases (Class 2.1) – Storage or Use	In a quantity of over 1 ton	In a quantity of 100–1,000 kg	In a quantity of under 100 kg
Non-Combustible and Non-Toxic Gases (Class 2.2) – Storage or Use			For every quantity
Toxic Gases (Class 2.3) – Storage or Use	In a quantity of over 100 kg	In a quantity of under 100 kg	

Source: Hazardous Substances Regulations, http://www.sviva.gov.il/English/Legislation/Documents/ Hazardous%20Substances%20Laws%20and%20Regulations/HazardousSubstancesRegulations- CriteriaForDeterminingPermitValidityPeriod-2003.pdf.

Anhydrous ammonia, in a quantity of over 5 tons, would be regulated as Level A, which would correspond to a one-year permit length. The "Business Types" referred to above are also contained in Table A. Among these are "Refineries and Petrochemical Industries," "Textile Plant [sic]," and "Hospital, Clinic or other Medical Institution."[364] The "Activity Types" are in Table A as well, with types such as "Metal Plating, including Galvanization, Printed Circles, Surface Treatments for Metals and Anodization," "Pesticide Storage," and "Sewage Treatment, including Neutralization."[365]

Looking at Table B of the Schedule, we may see classifications of the various UN Organization – "Recommendation on the Transport of Dangerous Goods: Model Regulations" hazard classes. It should be noted that where a substance has two hazard classes – a primary and a secondary – the determining hazard group shall be the stricter of the two.[366]

Poisons Permits may be granted for various periods of time, between one and three years. The differentiating factors for the terms are the type and quantity of the hazardous substance at issue, the type of industry or activity involved, and the potential environmental risk of the hazardous substance category.[367] These categories are based on risk, and broken down into three groups – A, B, and C – with A being the most restrictive level (e.g. permits issued annually, and the holder will be subject to more frequent inspections and higher fees than those in other categories).[368]

The Hazardous Substances Regulations (Classification and Exemption) No. 5756-1996 arose from Sections 10, 12, and 13 of the HSL, and define several categories of hazardous materials, as well as actions relating to their classification. Schedule One of the Hazardous Substances Regulations (Classification and Exemption), entitled "Regulation 1 – Hazardous Materials of Category 'A' and Hazardous Materials of Category 'B'" lists more than 225 substances, each with a corresponding concentration (expressed as a percentage) or quantity (expressed in kilograms). Of the more than 225 substances listed in Schedule One, 19 also appear in Schedule Two, and are exclusively pesticides (e.g. Number 154 – "Pesticide, dithiocarbamate"). Using the foregoing definitions, phenacetin is a "hazardous material," as it is listed in Schedule One as a "Hazardous Materials of Category 'A,'" when present in a concentration of less than or equal to 0.1 kg. As it is not present in the "Hazardous Materials of Category 'B'" list, phenacetin is not considered a hazardous material for that category. Bromine is also a "Hazardous Materials of Category 'A,'" at a concentration less than or equal to 1%, and also at a concentration less than or equal to 5 kg.[369]

The Definitions section, Section 1, addresses "Category A hazardous material," defined as "a hazardous material listed in the First Schedule in the concentration stated next to it, or in a lower concentration," and "Category B hazardous material," defined as "a hazardous material listed in the First Schedule in the quantity stated next to it, or in a smaller quantity." Interestingly, "Hazardous materials" themselves are defined, self-referentially, as "a Category A hazardous material, a Category B hazardous material and radioactive elements and their compounds."[370] The reader should note, however, that a more accurate writing of this definition would be to replace the words "and" with "or." As the text reads, it appears that to be a "hazardous material," the substance must, for example, be listed in both Category A <u>and</u> Category B <u>and</u> as a radioactive element <u>and</u> their compounds. These categories and types should be read severally, as well as potentially jointly to make more sense.

Notably, as per Section 2, the provisions of Sections 3, 4, 5, 8(1), and 9 of the HSL shall not apply to a "hazardous material," except as in Section 3. The relevant points of Section 3 are that the provisions of Section 2 shall not apply (1) "to any of the hazardous materials specified in Schedule Two in a total quantity of more than 50 kg" or (2) "to a dealer in forty or more Category B hazardous materials, excluding domestic use."[371]

Finally, the Pharmaceutical Ordinance of 1981, administered by the Institute for the Standardization and Control of Pharmaceuticals (Institute) under the Ministry of Health, is another component of Israel's chemical substance management infrastructure. The Institute, comprised of several analytical laboratories, is primarily responsible for quality assurance of pharmaceuticals marketed in Israel.[372] Note that the Institute has jurisdiction over the products <u>marketed</u> in Israel, irrespective of whether they are imported or produced in-country.[373] Registration is only granted by the MOH. The Institute has responsibility for the following areas:[374]

1. Evaluation and control of medicines intended for human and veterinary application, to ensure quality, safety, and efficiency of pharmaceuticals;
2. Evaluation of the safety of cosmetic products;
3. Evaluation of efficiency and safety testing for pesticides;
4. Toxicological and analytical evaluation of dossiers for human and animal drugs; and
5. Development and improvement of quality control methods for the measurement of pharmaceuticals.

In 2010, Israel set about the bid process for what was described as a "comparative analysis" among various international chemical management systems, such as the EU's REACH and the United States' Toxic Substances Control Act (TSCA), among others. The manifest goal of this process was to establish a new administrative unit which would have responsibility for chemical registration and assessment.

Pesticide Regulations

The Israeli pesticide registration process is fairly comprehensive, and not one into which entities should enter lightly:

The pesticide registration process begins with testing and investigation over a period of two years, following provisional approval for limited use. When comprehensive toxicological data have been gathered, an advisory committee, composed of representatives of several ministries, including Environmental Protection, Health, Industry, Trade and Labor, as well as representatives of consumers, decides whether to approve the product for final registration. Materials are assessed for their environmental impact, endurance, risk to groundwater and other factors. For registration, the Pesticides Division of the Plant Protection and Inspection Services has established criteria for submitting a toxicology file to the interministerial committee for coordination of pesticide use. The criteria are largely based on European directive 91/414/EC concerning the placing of plant protection products on the market.[375]

As with many other national pesticide regulations, Israel segments its registrations into various areas, based on the intended use of the pesticide product.

The registration process for pesticides for public health purposes is separate from that for agricultural use. Israel's regulations, approved in 1994, comply with strict international standards and require applications for the registration of a new molecule, new formulation, renewal and change of name/label/packaging, etc. The regulations require the officer responsible for pesticide registration to consult with an interministerial professional committee, composed of representatives of the Environmental Protection and Health Ministries. Applications must be accompanied by specification of the composition of the pesticide, [a] copy of the proposed label, toxicological file on the impact of the material on humans, the environment, flora and fauna, material safety data sheet, and more.[376]

The Plant Protection Law (1956):

grants the Minister of Agriculture authority, following consultation with an advisory interdisciplinary committee, to regulate the import, sale, distribution and packaging of pesticides, fertilizers and other materials. [Further] the law authorizes the Minister of Agriculture to regulate the use of pesticides, to require a permit for their use, to promulgate regulations on the safe use of pesticides and to prohibit or limit the use of pesticides deemed hazardous to human health and the environment.[377]

In 1994, the "Import and Sale of Chemical Preparations Regulation" was enacted, which "relate[s] to the registration of a process of a preparation that is manufactured or imported [for pesticidal use]." As with national schemes in Venezuela and China for other chemical products, which mandate that certain data be generated under conditions and/or on species specific to their country, data submitted must include "professional literature on the results of experiments with the preparation under Israeli conditions."[378]

Also in 1994, the "Hazardous Substances Regulations (Registration of Preparations for the Control of Pests Harmful to Humans)" (Preparations Regulation), 5754-1994 was issued under the authority of Sections 10(3), 12, and 17 of the HSL. The Preparations Regulation was amended in 2002. This "prohibit[s] the production, import or maintenance of any substance that is not registered and that does not comply with the conditions stipulated in the registration certificate and in the regulations."[379] While readers familiar with global chemical substance regulations may reflexively identify "preparations" as a mixture or solution composed of two or more substances, the Preparations Regulation specifies preparations whose purpose is expressly defined as those which control pests deemed harmful to humans.[380] Indeed, Section 1 defines the term as:

> A material which is hazardous in any state of aggregation, or mixtures thereof, intended to be used for extermination or repulsion of insects and other types of arthropods, as well as rodents and other types of vertebrates that are a nuisance, inflict damage or might inflict damage to humans or their property; with the exception of any item or substance intended to be used on the human body.[381]

Section 2 states that no person shall sell, "hold for the purpose of selling," manufacture or import, or use a preparation unless that preparation has been registered, and then only according to the conditions granted under the registration process.[382] Furthermore, Section 2(d) prohibits the use of a preparation outside of the "instructions of use that are specified on its label, and in accordance with its purpose of use."[383] Readers familiar with global pesticide regulations may recognize this as language similar to the United States' FIFRA regulation, which expressly declares that for such products, "It is a violation of Federal law to use this product in a manner inconsistent with its labeling."[384]

The authority for registering pesticides is delegated to an officer appointed by the Environmental Protection Minister, known as the "Supervisor." Although the Preparations Regulation prescribes a multitude of requirements, the Supervisor is given a relatively large degree of flexibility and deference to mandate additional requirements. As per Section 2A(a), for the production or import of unregistered pesticides, the aspirant must apply using the form provided in the Preparation, as well as the required documentation. Section 3(a) requires the applicant to provide documentation, in the form of appendices to the application, as follows. Twelve copies of each item should be submitted, in Hebrew, unless they are written in English.[385]

1. Results of experiments with the preparation and literature regarding its mode of use, which testify to the preparation's efficiency in the fulfillment of its intended goals;
2. (Omitted in 2002 Amendment);
3. In case of an imported preparation, documents testifying to the preparation's registration and sales in other countries, including the labels in these countries;
4. If required by the Supervisor, [a] sample of the preparation's packaging;
5. A detailed summary of the toxicological file and Safety Data Sheets of the preparation, its active substance and each of its components;
6. If required by the Supervisor, a complete toxicological file, including acute and chronic toxicity data in respect of the preparation's impact on humans, the environment, animals and plants;
7. Information on the preparation's shelf life;
8. (Omitted in 2002 Amendment);
9. A quality check certificate from the laboratory;
10. A copy of the proposed label said in regulation 5; and
11. If required by the Supervisor, additional data about the preparation.

Registration is valid for a period of either three or six years from the date of issuance, or for a shorter period as determined by the Supervisor, as per Section 10(b). Renewal instructions are detailed in Section 12(a) of the Preparations Regulation, and the penalties for violating the terms of registration, or for advertising and/or selling an unregistered preparation, are set out in Section 17(a).

Packaging containing the preparation must, under Section 5(a), include a label attached to or printed on it, "in a manner that prevents removal," unless the Supervisor decides to attach the label via a different method.[386] The data printed on the label, in "clearly visible letters," in both Hebrew and Arabic, should include:[387]

1. Manufacturer's name and address;
2. Importer's name and address;
3. Preparation's commercial name;
4. Preparation's formulation;
5. Common names of all active and synergistic substances;
6. Concentrations of the preparation's active and synergistic substances;
7. Preparation's purposes of use;
8. Net weight or volume of the preparation in its package using the metric system;
9. Preparation's production batch number and expiry date, or preparation's manufacture date and expiration period;
10. If required by the Supervisor, [the] preparation's classification;
11. Preparation's mode of use;
12. Detailed precautions in respect of the preparation's storage and use;
13. Registration number of the Ministry of Environmental Protection and its year of expiration;
14. Instruction to see a doctor in case of poisoning suspicion;
15. Poison and flammability signs, as specified in regulation 8; and
16. The words "The Ministry of Environmental Protection has determined that use of this preparation contrary to the label instructions, [sic] might endanger the health of the public, the user and the quality of the environment."

The Supervisor may require the applicant to add more information to the preparation's label.

A "poison sign" must also be printed, on a diamond-shaped background and in black on a white background, on the label, as per Section 8(a). The poison sign shall include a "skull and crossbones" image, with the word "Poison" printed in Hebrew, Arabic, and English. The minimum length of each side of the diamond is 15 mm, and the marking should be "clear, easily readable and permanent."[388] The poison sign must appear on every label of the preparation, unless the Supervisor directs otherwise. Additionally, the Supervisor may require a flammability sign, or any other sign, at his/her discretion, as set forth in Section 8(c).

The Ministry of Health and the Ministry of Environmental Protection, as part of a committee composed of an equal number of representatives (three) from each Ministry, reviews all pesticide applications and advises the appointed officer, as per Section 9(a).[389] "The regulations stipulate a number of cases in which the appointed officer, with the agreement of the… committee, may refuse to register a particular pesticide… may cancel registration or make [registration] conditional upon the fulfillment of a number of prerequisites."[390] Examples of a situation in which registration may be refused include if: (1) another pesticide, which is less harmful to humans or the environment, is already registered for the same intended use, or (2) the applicant did not provide proof, to the satisfaction of the committee referenced above, that the preparation is effective in the use scenario(s) described in the application materials.

In a situation where one person holds a registration for a preparation, and another person seeks to register the same preparation under his/her name, the additional registrant must submit the form, along with the following supplementary information:[391]

1. A copy of the original registration certificate;
2. Written consent of the registration certificate holder to transfer the registration from his name to the applicant, and to allow the applicant to rely on his original application;
3. In case of an imported preparation – approval of the preparation's manufacturer regarding his consent to transfer the registration of the preparation from the holder of the registration certificate to the applicant;
4. A quality check certificate from the laboratory;
5. A copy of the label that was approved in the registration certificate and of the proposed label referred to in regulation 5; and
6. If required by the Supervisor, additional data about the preparation and the transfer of registration.

Upon approval of what Section 2B(b) calls the "requested transfer of registration," although this is a term which is known to have a different meaning in other global registration regulations, the Supervisor will provide the additional registrant with a registration certificate in the additional registrant's name, whose expiration date will be the same as that on the original registration certificate.[392]

Finally, the 1964 "Workers with Pesticides Regulation" (Workers Regulation), which derived from the 1946 Work Safety Law, "seek[s] to ensure the safety of those working with pesticides… prescribe[s] conditions for the storage of pesticides, handling of pesticide packages and preparations for pest control, protective gear for… workers, care of pest control devices, and responsibilities of pest control workers."[393]

Occupational Safety and Health Regulations

OSH regulations in Israel date from 1940, with the passage of the Public Health Ordinance (Ordinance). The Ordinance sets out the powers of the Ministry of Health and the Ministry of Environmental Protection which may be used to control public health and to manage various types of environmental nuisances.[394] Presently, OSH is managed under the Ministry of Economy and Industry.

The Safety and Operational Health Administration (SOHA) was created through implementation of both the Labor Inspection (Organization) Law, 5714-1954 and the Work Safety Ordinance, 5730-1970.[395] The manifest purpose of SOHA is to prevent work-related accidents and maintain the health of all employees in Israel.[396] "The Administration is therefore responsible for defining policies for work safety, employee health, and occupational hygiene, and for supervising and enforcing the various related laws and regulations at workplaces in Israel."[397]

While there are a multitude of occupation- and industry-specific OSH regulations in Israel [e.g. Safety at Work Regulations (Occupational Hygiene and Health of the

Public and Workers with Harmful Dust), 5744-1984; Safety at Work Regulations (Environmental and Biological Monitoring of Workers with Hazards), 2011], the Worker Safety Ordinance of 1970, its 1982 Amendment No. 2 and its 2016 Amendment No. 9, are perhaps the most overarching. When enacted into national law, the competent authority was the Ministry of Labor and Social Affairs, the present administering agency is the Ministry of Economy and Industry.

Chapter 1, Article 1, Section 2 describes those to whom the Work Safety Ordinance is applicable. As one might expect with a regulation that was developed over 45 years ago, and that was one of the first in its field, the terminology is perhaps necessarily broad, at least when compared with current legislation. Article 1, Section 1 describes the scope of the regulation generally, using the word "undertaking."

> The expression "undertaking" means any premises in which, or within the close or precincts of which, persons are engaged in manual labor in any process for or incidental to the making, altering, repairing, ornamenting, finishing, cleaning, washing, breaking up, demolition, or adapting for sale, of any article or part of any article and in respect of which the following two obtain:
>
> (1) the work of the undertaking is carried on by way of trade or for purposes of gain;
> (2) if hired workers are employed therein, their employer has the right of access or control.[398]

The bulk of the 1970 Work Safety Ordinance is again fairly general in its scope and direction. It contains such items as "Cleanliness and Painting" (Chapter 2, Article 1), "Overcrowding" (Chapter 2, Article 2), "Ventilation, Lighting and Temperature" (Chapter 2, Article 3), and "Drainage of Floors" (Chapter 2, Article 4).[399] Chapter 2, Article 5 ("Medical Supervision") does contain some of the language and conditions seen in other chemical substance-related workplace safety and health statutes. Specifically, it empowers the Minister to enact regulations requiring "such reasonable arrangements to be made as may be specified therein for the medical supervision of the persons, or any class of persons, employed in any undertaking or class description of undertaking if" one of three defined aspects applies.[400] These are:[401]

(1) There have occurred, or there are reasonable grounds for apprehending that there may occur cases of illness attributable to any work, material or process in the undertaking;
(2) By reason of changes in any process or in the substances used in any process, there may be a risk of injury to the health of persons employed in that process; and
(3) By reason of the introduction of any new process or new substance for use in a process.

The remaining Articles of the Work Safety Ordinance are devoted to requirements for specific aspects and work environments, such as "Lifts," "Steam Boilers," "Air Receivers," and "Safety Provisions in Case of Fire."[402]

Additionally, the Israel Institute for Occupational Safety and Hygiene (IIOSH), a public non-profit organization whose stated goal is "promoting the conditions of occupational safety and hygiene," was established in 1954 under the Labor Inspection (Organization) Law. The IIOSH seeks to foster a culture of safety and health, and to promote "all that is related to safety and health in the workplace and through the life cycle."[403] The IIOSH website (https://www.osh.org.il/eng/main/) is a resource for updated information on new draft regulations which address worker and workplace safety.

Waste Regulations

Israeli waste regulations were developed under the umbrella of the Public Health Ordinance of 1940, specifically Section 62B, as well as under Section 10 of the Licensing of Businesses Law. Perhaps the most significant of these, the "Disposal of Hazardous Substances Waste Regulation" of 1990 (Waste Regulation), relates to specific means for identifying and disposing of wastes determined to be "hazardous." The Waste Regulation defines a "hazardous substance" in Section 1 as "a substance in any state of aggregation that has a U.N. number, as specified in the Orange Book, and as described in Part A of the First Appendix to the Inspection of Goods and Services Order (Transport and Trailer Services), 5739-1978," similar to how they are defined or referenced under other Israeli regulations (cf. the Hazardous [Industrial] Plants Regulation).[404]

Section 2(a) of the Waste Regulation decrees that a plant owner is required to dispose of all waste which either "originates in a plant or is found therein" as "soon as possible and not later than six month [sic] from the time of its generation."[405] Industrial (read here as non-hazardous) waste should be sent to the "plant" for neutralization and treatment, while hazardous substances should be packed and transported in accordance with applicable provisions and sent to the "Hazardous Waste Site" in Ramat Hovav.[406] The related "invoices of the Hazardous Waste Site or the place of recycling or reuse," demonstrating proper disposal, must be maintained and preserved by the plant owner, as per Section 3, and shall be submitted to the Director, the Licensing Authority, or anyone acting on their behalf, upon request.[407]

Related in concept, but not directly in regulatory lineage, is the 1994 "Import and Export of Hazardous Substances Waste" (Import/Export). The Import/Export regulation derives from Sections 10, 12, 13, and 17 of the HSL, but not from the Licensing of Businesses Law, as has been seen previously. The requirements of the Import/Export regulation are clearly set out in Section 2, declaring:

> A person shall not import to Israel or export from it hazardous substance waste [defined in Section 1 as "material of any kind or form that contains a hazardous substance as defined in the (HSL)"] except in accordance with a permit certificate… after it is proven to the satisfaction of the Supervisor [defined in Section 1 as "the person that the Minister of Environmental Protection authorized for the purpose of all or part of these regulations"] that: (1) the import or export is done from a country that is party to a convention (defined as the 1989 Basel Convention on the Control of Transboundary Movements of Hazardous Wastes and their Disposal) or to such a country, as the case may be, and (2) for import, all of the following conditions are met:[408]

> 1. The hazardous substances waste is intended for recovery;
> 2. The person applying for the permit has the information required regarding the type of hazardous substances waste and its composition;

3. The import of hazardous substances waste to Israel, its transfer, storage, maintenance, use and handling do not endanger public health or the environment in accordance with the precautionary principle; and

4. The import of hazardous substances waste to Israel, its transfer, storage, maintenance, use and handling is done in a manner that does not cause harm to public health or environmental quality.

Permits to import hazardous substances waste into Israel must be reflective of a bank guarantee or other suitable bond, specified by the Supervisor, that will serve to ensure compliance with the permit, or third-party liability insurance by an insurer licensed to operate in Israel. Such a guarantee or bond may be confiscated for a breach of any of the conditions of the permit, provided that the permit holder has "suitable opportunity" to argue against the forfeiture.[409]

With respect to the export of hazardous substance waste, the competent authority in the target country must consent in writing "to receive into its jurisdiction the hazardous substances waste being exported, and the export is carried out in accordance with the established conditions or requirements and in accordance with the convention."[410] Both the import and/or export volume of hazardous substance wastes are required to be reported upon the request of the Supervisor, who in turn shall report the same annually to the Internal Affairs and Environmental Protection Committee of the Knesset.

Safety Data Sheets and Labels

The Israeli Standard (IS) for "Labeling and Marking of Dangerous Substances" (IS 2302) ("Classification, Packaging, Labeling and Marking Dangerous Substances and Preparations") (revised in 2009) presents a mandatory harmonized system that regulates the labeling of chemicals (partly correlating with the former EEC Dangerous Substances Directive 67/548/EEC). IS 2302 delineates labeling requirements for chemicals in-house, during transport from facility to facility within the same compound, and during the transport of dangerous substances by road or rail. More specifically, "the standard classifies dangerous materials according to hazard groups, packaging size, [sic] and packaging type, and outlines requirements for marketing, packaging and labeling of these substances."[411] Here, the risk and safety phrases are provided in Hebrew, and are based on the UN's "Orange Book" definitions.

The 1998 Work Safety Regulations ("Safety Data Sheets, Classification, Packing, Labeling and Marking of Packages"), 5758-1998 (amended November 8, 2011), created under the Safety at Work Ordinance 5730-1970, and Sections 10(2), 12, and 13 of the HSL:

> Require[s] producers, importers, distributors or sellers of a hazardous substance to supply recipients with Material Safety Data Sheets (MSDS), and call for the maintenance of an MSDS in the factory or business in order to inform users about hazards in their workplace. The MSDS compiles relevant data on a hazardous substance in a uniform format and provides information on how the material can be safely handled, used and stored.[412]

It should be noted that a SDS is only needed for substances which fall under the definition of "poisonous substances." Substances which are hazardous, but not poisonous, can be sold without an SDS.[413]

As laid out in Section 2 of the Work Safety Regulations, "the purpose of these regulations is to safeguard the safety and health of people working with hazardous substances and those in the vicinity thereof, and the environment."[414] Section 7(a) directs that the Work Safety Regulations do not apply, however, to (1) hazardous substances in a quantity or concentration which is exempt from the provisions of Section 2 of the Hazardous Substances Regulations (Classification and Exemption) 5756-1996; (2) food, as defined in the Public Health Ordinance 5743-1983; (3) "Medicaments" and "Medicinal Drugs," as defined in the Pharmacists' Ordinance 5741-1981; and (4) products that contain hazardous substances, as defined in Chapter G1 of the Pharmacists' Ordinance 5741-1981 and are sold to the public for home use.[415]

As per Section 3(a), a manufacturer, importer, dealer, or distributor of a hazardous substance ("hazardous substance" is defined here in the same way as under the HSL) must pack the substance, or ensure that it is packed and marked within the parameters of the Work Safety Regulations, and that a SDS is attached to it.[416] Further, a "holder of a workplace," set out in Section 1 as any of the following: the workplace owner, the acting manager of the workplace, the person under whose supervision and inspection the workplace operates, or (if the industrial plant is owned by a corporation) the acting manager of the corporation, shall ensure that a SDS for every hazardous substance used in the workplace is "readily available" for workers.[417] Finally, those who are tasked with receiving hazardous substances should ensure that such is accompanied by a SDS, and the "manufacturer, dealer, importer or distributor" of a hazardous substance shall provide a SDS upon request.

Section 4(a) mandates that the SDS for a hazardous substance shall be filled out in accordance with the guidelines [emphasis the author's] in the Schedule at the end of the Work Safety Regulations, shall be based on the most up-to-date information, shall be written in Hebrew or English, and shall contain the following information:[418]

1. Identification of the hazardous substance and identity of the manufacturer, importer, dealer, or distributor, as the case may be;
2. Identification of the hazardous substance's components;
3. The hazardous substance's risks;
4. First aid instructions;
5. Fire extinguishing procedure;
6. Precautions;
7. Handling and storage;
8. Means to minimize exposure, and Personal Protective Equipment (PPE);
9. Chemical and physical attributes;
10. Stability and reactivity;
11. Toxicity (toxicological data);
12. Environmental data;
13. Disposal methods for a hazardous substance;
14. Transport;
15. Legislation and standardization; and
16. Other information.

The statement "The information contained in this sheet is based on the best knowledge and experience currently available" must appear at the end of every SDS, as per Section 4(e).[419] The SDS should be updated when there is "essential new information" regarding a hazardous substance which may impact the safety and health of those handling the material, or on the environment, as per Section 5. Also, in such a situation, the updated SDS should be provided to "every person who handles hazardous substances and that had received from him such a substance during the twelve months preceding the update."[420]

Section 4(d) addresses the issue of Confidential Business Information (CBI) with respect to information provided on the SDS. Specifically:

> it is possible not to deliver information which is regarded as trade secret or professional secret [although the terms "trade secret" and "professional secret" are neither defined nor referenced in the Work Safety Regulation], providing that the information that is delivered enables safe handling of the substance, to ensure protection of the well being [sic] of workers in the workplace, of safety and environmental quality.[421]

As per the Work Safety Regulations, the question of confidentiality of raw materials used is addressed somewhat obliquely; indeed, "the concentration range of impurities should be indicated alongside the definition of the substance. However, the concentration of the ingredients may be confidential, even if only the concentration range is stated rather than the exact concentration in the mixture."[422] This likely stems from the point that, under Israeli law, the manufacturer is required to provide the CAS number of substances listed on the sheet. Specifically, "there is no obligation to provide information on the ingredients of a preparation for which a safety data sheet must be supplied. However, a description of impurities with their concentration range must be provided."[423]

> Furthermore:
> if the data is a trade secret it can be kept secret provided the relevant data regarding the safety of workers and environment is made public. A manufacturer may make use of this clause only if it can be argued that the information relating to a commercial secret:[424]

- Is not accessible by the general public;
- Cannot easily be detected by others;
- Gives an added advantage to the owner;
- Is kept secret through steps taken by the owner.

As noted above, the Schedule at the end of the Work Safety Regulations contains guidance as to the information to be included on the SDS. Regulation 4(a) under the Schedule, "Guidelines for the Preparation of a Safety Data Sheet," discusses six items which should be taken into account when creating the document:[425]

1. This Schedule specifies only the main particulars to be filled into the Safety Data Sheet and the person filling out the Sheet may add particulars, at his discretion [as observed earlier, this section does not include any mandatory language, such as "shall" or "must"];

2. All the chapters and parts of the Safety Data Sheet, as specified below, must be filled out; no spaces should be left empty spaces [sic], nor should one write, unjustifiably, "irrelevant" or "no available information";

3. Information which has no explicit connection to any of the Sheet's chapters shall be written in Section 16 ("other information") of Chapter B;

4. The chapters of the Safety Data Sheets shall be filled briefly and clearly, according to the guidelines of this Schedule;

5. The information in the Safety Data Sheet should match the information written on the pack's label – if such exists; and

6. The date at which the Safety Data Sheet was prepared or its last update shall be written at the top of the Sheet.

Item 7 in the list provides potential references for creating the SDS (e.g. "The most updated ANSI Z400.1-1993 Standard"), while Item 8 notes that the potential references are available in the Information and Response Center of the Ministry of Environmental Protection in Ramla, as well as in other locations.[426]

Chapter B of the Schedule, "Filling out the Sheet's Chapters," presents detailed guidance on information to be included in each section. As the data is extensive, it will not be specifically examined here, other than to note that the information in Chapter B is generally similar to information required under the ANSI Z400.1 format in the United States, which was widely used prior to the U.S. implementation of the Globally Harmonized System of Classification and Labeling (GHS) through the Hazard Communication Standard, issued by the Occupational Safety and Health Administration (OSHA) in 2012.

With respect to the GHS, Israel issued Standard SI 2302 in 2004, with an update in 2009 and in 2014. Part 1 of the Standard proposed to adopt EU Directive 67/548/EEC and 1999/45/EC, relating to dangerous substances and preparations, sorting, packaging, and labeling, while Part 2 proposed to address the transportation of these same substances and preparations, using the UN "Orange Book."

> Israel notified the World Trade Organization on 3 December 2013 of a draft revision of standard SI 2302 parts 1 "Dangerous substances and preparations: Classification, packaging, labelling and marking" and 2 "Dangerous substances and preparations: Transportation, classification, packaging, labelling and marking" which will implement the GHS in Israel. Public consultation was closed on February 2014. Comments received during the consultation period are under consideration.[427]

To date, the GHS has not been implemented in Israel, and there are no clear plans to do so.

5

Kuwait

National Overview

The State of Kuwait (Kuwait), formally established as a sheikhdom in 1756 by the Āl Ṣabāḥ family, is situated in the northeast of the Arabian Peninsula in Western Asia.[428] From 1899 to 1961, it was a British protectorate.[429] Although in 1961 Kuwait became the first of the Persian Gulf Arab countries to gain independence, its governmental history has been somewhat volatile.

> Since coming to power in 2006, the Amir has dissolved the National Assembly on seven occasions (the Constitutional Court annulled the Assembly in June 2012 and again in June 2013) and shuffled the cabinet over a dozen times, usually citing political stagnation and gridlock between the legislature and the government.[430]

The official language is Arabic, and Islam is the official religion.

Governmental Structure

Kuwait, which gained independence from the United Kingdom on June 19, 1961, is a constitutional emirate with a parliamentary system of government. Its constitution combines aspects of both presidential and parliamentary systems of government.[431] Kuwait organizes its government into three distinct branches: Executive, Legislative, and Judicial. As part of the Executive Branch, the head of government is the Prime Minister. The Emir is a hereditary position, and it is he who appoints the Prime Minister and his deputies.[432] The Branch also encompasses a First Deputy Prime Minister and three Deputy Prime Ministers. The Prime Minister also appoints a Cabinet/Council of Ministers which is approved by the Emir. Under the Legislative Branch, "Kuwait's National Assembly has 66 seats, of which 50 are elected by popular vote and 16 cabinet ministers are appointed by the Prime Minster. Elected members serve four years."[433] Under its Judicial Branch, Kuwait employs a civil law system, with an independent judiciary. Sharia law is "significantly used for personal matters."[434] In each of the country's six governorates, "there is a summary court. There is also a court of appeals, a Cassation Court [a form of Appellate Court], and a Constitutional Court."[435]

Kuwaiti regulations are published in the Official Gazette of the State of Kuwait, *Al Kuwait al-Youm* (available at http://kuwaitalyawm.media.gov.kw/).

Chemical Regulation in the Middle East, First Edition. Michael S. Wenk.
© 2018 John Wiley & Sons Ltd. Published 2018 by John Wiley & Sons Ltd.

Key Chemical Regulatory Agencies

The Environment Public Authority (EPA) was established in 1995 by Article 2 of Law No. 21 (amended in 1996 by Law No. 16).[436] The EPA is considered a "public authority" under the Judicial Branch, with budgetary responsibility for environmental affairs, over which it has jurisdiction. It falls under the Council of Ministers, and under the supervision of the Supreme Council of Environment.[437] The First Deputy Prime Minister and Minister of Foreign Affairs is the Chairman of the Supreme Council for the Environment (Supreme Council).[438]

The Supreme Council is comprised of the following considerable number of representatives:[439]

- The Minister of State for Cabinet Affairs and Minister of Information;
- The Minister of Social Affairs and Labor and Minister of State for Economic Affairs;
- The Minister of State for Housing Affairs and Minister of State for Services;
- The Minister of Health;
- The Minister of Public Works;
- The Minister of Oil and Ministry of Electricity and Water;
- The Minister of Awqaf and Islamic Affairs and Minister of State for Municipal Affairs – General Authority for Agriculture and Fisheries Affairs;
- The Head of the Voluntary Work Center;
- The Member of the Supreme Council; and
- The Chairman of the Board – Director General of the General Authority for the Environment – Rapporteur of the Supreme Council for the Environment.

Article 4 of Law No. 21 gives the Supreme Council the opportunity to include three members with particular "expertise and efficiency" relating to the environment to a three-year term, which may be renewed for a similar term multiple times.[440] Article 5 directs that the EPA is to be headed by a Director General, "a specialist in the field of the environment," with a term and a renewal process in kind to that of the three members of the Supreme Council. Finally, as per Article 6, the Supreme Council will issue a resolution constituting a Board of Directors. Similar to the three members in Article 4, the Board of Directors will be comprised of eight "professional members of expertise and competence in the field of the environment, duly nominated from outside the Authority and appointed for a three-year term of office, renewable for similar terms."[441]

The EPA consists of ten departments, as follows:

- Industrial Environment Department;
- Planning and Environmental Impact Assessment Department;
- Administrative and Training Department;
- Coastal and Desertification Monitoring Department;
- Water Pollution Monitoring Department;
- Air Quality Monitoring Department;
- Administrative Affairs Department;
- Legal Affairs Department;

- Engineering Affairs Department; and
- International Affairs Department.

The EPA's website may be found at https://www.epa.org.kw/index.php, although it is relatively sparse in its level of content.

Some of the key areas in which the EPA has involvement and/or authority are:[442]

1. Setting and applying the general policy to protect the environment and setting strategies and work plans;
2. Defining lists of pollutants and setting standards for environmental quality, as well as preparing drafts of law, by-laws and requirements concerning environmental protection;
3. Setting the general framework for programs of environmental education and awareness, and leveraging this level of awareness to achieve increased environmental protection; and
4. Setting standards and requirements for project and establishment owners, and enforcing penalties against violators of these standards and terms.

Key Chemical Substance Regulations

In 1987, Kuwait implemented a national monitoring system for chemicals. It has a national committee on pesticides and ozone-depleting chemicals. This committee has been in service since 1990, and includes 16 concerned parties from Ministries and Non-Governmental Organizations (NGO), as well as a secretariat that is associated with the Ministry of Environment. Kuwait has been planning executive legislation on chemicals management, and has ratified all chemicals-related agreements including the Basel, Montreal, Stockholm, and Rotterdam Conventions. Additionally, the country has plans to implement a customs system for chemicals management. It is expected that the GHS, whenever implemented, will facilitate this effort. Challenges are foreseen, however, as the number of chemicals transactions that cannot be adequately controlled continues to increase, and unregulated storage of chemicals is a problem. There also continues to be a general lack of awareness and resources around chemicals management.[443]

Law No. 21 of 1995 "Establishing the Public Authority for the Environment" (Law No. 21), amended by Law No. 16 of 1996, is also known as the "main environmental law" within Kuwait. In addition to establishing the Kuwait Environment Public Authority under Article 2, as discussed earlier, Law No. 21 also gave the EPA the ability to follow up on or evaluate (environmental) impact assessment studies of projects, to implement the "polluter pays" principle, to promote optimal use of oil resources, and to demand consumption reductions and energy-saving technology in project design, etc.[444] The 1996 Amendment, Law No. 16, clarified the EPA's role relative to conservation, protection, and liability concepts.[445]

Law No. 21 itself contains 89 Articles and 20 Appendices, grouped in the following manner:[446]

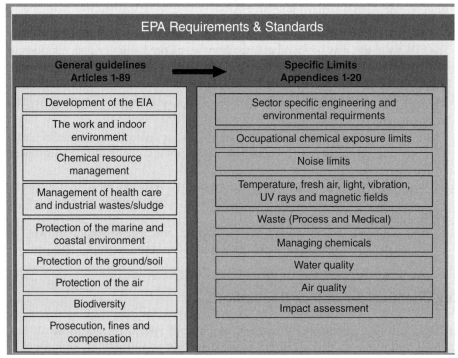

Source: American Society of Safety Engineers.

Key among the provisions of Law No. 21 are Article 10, which expressly allows the Board of Directors to suspend the operations of "any firm or activity or prohibit the use of any tool or material, whether completely or partially, if the continuation of the use of such tool and/or material shall result in environmental pollution," for a period of one week, which may be extended an additional week, and may require the implementation of precautionary measures during this period, and Article 18, which makes the provisions of the Law applicable to "all private and public sectors authorities and parties."[447] For any further extension of the suspension period, a judicial writ from the Chief Justice of the Court of First Instance is required. For readers not familiar with the term, in general, a court of first instance is a trial court with original or primary jurisdiction. The Director General may also issue a suspension order, in an emergency period, for a maximum of seven days.[448]

Additional Articles of Law No. 21 which may impact chemical substances are:

- Article 11 empowers the Director General, or any other competent authority, to delegate staff to inspect facilities where violations have been observed. Their actions may include the creation of records, taking samples and conducting necessary studies or measurements, and ensuring that any precautionary measures set forth under Article 10 are implemented. Interestingly, such staff "shall have the capacity of judicial police," and may involve the police as they deem necessary.[449]
- As per Article 12, the Board of Directors may request any documentation it desires from entities causing an environmental issue, as well as establishing laboratories and/or control plants which are "specialized in giving the final opinion concerning the outcome of laboratory tests relating to environmental pollution."[450]

- Article 13 sets out the penalties for contravention of Articles 8 and/or 10. Specifically, the violator shall be punished with a period of imprisonment not to exceed three years, or a fine of not more than 10,000 Kuwaiti Dinars.[451] It should be noted, however, that such penalties do not preclude involvement of the Kuwaiti court system. The court "may render judgement confiscating the things which caused the environmental pollution or damage, besides holding the parties responsible for the same liable to pay all necessary costs for rectification of the environmental damage and harm incurred…"[452] Furthermore, any individual(s) who "prevents the competent officers from carrying out their duties [recall from Article 11 that such "competent officers" could be inspection staff appointed by the Director General]… shall be punished with a term of imprisonment not exceeding one year and a fine of not more than 1,000 Kuwaiti Dinars."[453]

Articles 15 and 16 relate to the budgetary responsibilities of the Public Authority for the Environment (e.g. forecasting, allocation), while Article 17 provides for the issuance of the "necessary resolutions, rules and regulations for the purpose of operation of this Law."[454]

On October 2, 2001, the EPA issued "Decision No. 210/2001 Pertaining to the Executive By-Law of the Law of Environment Public Authority" (Decision No. 210). Entities, defined in Article 2 as "all governmental, joint, private parties and others," undertaking any of the activities listed in Appendix No. 1 of Decision No. 210/2001, "Development and Environment – The Environmental Impact of the Developmental Projects Project List" [sic], must submit a plan to the EPA. These activities include:[455]

1. Natural resources projects of fossil origin (e.g. drilling projects);
2. Natural resources projects of non-fossil origin (e.g. sand projects; aluminum fusion, manufacture, and storage projects; manufacture and storage of cement);
3. Other industrial projects (e.g. "manufacture, filling and storage of all the chemical projects," "manufacture, filling and storage of all pesticides projects");
4. Communication projects (e.g. communication and transmission towers erection);
5. Food, animal, and agricultural resources projects (e.g. fish farming, tanning, and manufacture of animal skins);
6. Projects related to housing and industry;
7. Projects of special nature (e.g. hospitals and health facilities, military projects);
8. Waste Appendix No. (1) related projects (e.g. treating waste projects);
9. Power generation and water desalination; and
10. Land, air, and marine transport (e.g. expressways, railways, tunnels).

The requirements for such a plan are ambitious. Article 4 requires the party/parties to submit an initial report to the EPA, which will "study and to [sic] give its opinion" within 60 days.[456] Such a report should include the following items:[457]

1. A complete technical description of the specified project, or the modification or expansions which are proposed to be introduced to an existing project, together with the necessary engineering plans, type of technology, equipments [sic], means and materials which would be used in the building or the expansion;
2. A statement of the economic and social feasibility of the suggested project;
3. A comprehensive description of the environmental project and the surrounding areas, which may be affected by the project execution, or introduction of modifications or expansion to an existing project;

4. A comprehensive statement on the expected impacts on environment as a result of the proposed project execution;

5. An evaluation of negative, positive, accumulative and non-accumulative, direct and indirect impacts on the short-term and long-term stages, on the environment during the various phases of the project execution (as from the preparation stage, execution, operation, maintenance, accomplishment till after the expected age of the project or cancellation thereof), as well as the scientific illustration applied in assessment of these effects;

6. An overall statement of the steps that should be brought about in order to restrain or reduce the negative effects of the project on environment, which may be exposed to harm on the short and long run; and

7. A commitment of [sic] applying continuous protection measures after project accomplishment, with necessary monitor and control systems that must be followed.

After obtaining the EPA's approval on the initial report, the applicant(s) should work with the EPA to determine the time needed to prepare and submit the final report, which will include the environmental impact of the proposed project. As before, the EPA has 60 days to review the final report and give its opinion.[458] Additionally, Article 5 allows the EPA to ask for desired information, statements, or documents, or to require the carrying out of additional studies related to the environmental impacts on the suggested project prior to presenting the final report.[459]

Chapter III, "Chemical Resources Management," is perhaps the section most directed toward chemical substance management. In fact, the section's subtitle – "The environmental criteria for chemical substances production, safety, transport, storage, import and export from and to the State of Kuwait, and the customs transit through its territories" – brings this aspect home quite clearly. Article 14, the first Article in this Section, as the Article numbers carry forward from previous Chapters, defines those to which the Chapter is applicable:

> All parties which produce, fill, handle, transport, import, export and deal with customs transit of chemicals should abide by the environmental conditions and criteria stipulated herein, and comply with the classification of dangerous chemicals mentioned in Appendix No. (10-1).[460]

As per Article 15, any party who "would produce, import or export chemicals" must obtain a license from the "concerned authorities." Such licensing authority may cancel the license if the product is shown to be harmful to the environment or human health.[461]

Beginning with Article 16, Chapter III details the requirements applicable to those who produce, import, or export chemical materials (the Parties). Under Article 16, "The parties who produce, export and import chemicals should maintain a record numbered and sealed by [the] Environment Public Authority containing the following information:"

(1) Type and quantity of the raw materials used in production.

(2) Type and quantity of the chemical product.

(3) Any other details specified by the Environment Public Authority.[462]

Article 17 discusses the requirements for the Parties to observe when refilling containers with chemical product. Specifically, the Article covers:

1. The container requirements ("The refill should be of good quality from inside that suit[s] the substance inside it, and may not be affected by acids, alkaline, and solutions. The refill must be painted with a substance resistant to rust, erosion, and reaction. It should be tightly closed, not to be fragile and can bear all transport circulation, vibration, and thermal changes circumstances.").[463]
2. The labeling requirements, including the adequate size of the container to contain applicable labeling ("The volume of the refill must be suitable to contain all signs, information, pictures, drawings, and symbols internationally recognized and which show dangers [of] toxicity of these substances, how they can be opened emptied, used or disposed thereof. All these details must be placed clearly on the refill, and details cannot be removed, or modified according to the instructions listed in the Appendix Nos. (10-2, 10-3) attached with this by-law.").[464] The text should be written in Arabic, and should contain the following data:[465]
   - Name of the manufacturing company, production and expiry dates, operation and registration numbers;
   - Refill content, chemical and trade names, activity substance, total and net weights, concentration degree, type of danger and toxicity;
   - Steps to be taken in emergency cases that may cause harm to the environment and public hygiene; and
   - The appropriate storage.

Further, Article 17 gives specific requirements for those entities that import and/or export chemical substances. Section (iv)(1) requires the following details to be submitted to the EPA to import or export chemical substances:[466]

1. A list of ingredients;
2. The ["]serial number["] of the substance;
3. Health and environmental impacts;
4. Purpose, the importing or the exporting party;
5. Precautions that should be applied upon emergency cases;
6. Chemical and physical specifications;
7. Product classification number or the customs statistical number according to the organizing system; and
8. Ideal method of substances discharge or their containers.

Additionally, the importing and exporting Parties of chemicals should abide by the provisions of prior approval agreements (e.g. Prior Informed Consent under the Rotterdam Convention), as well as other international agreements effective in Kuwait.

Finally, Article 18 relates to the requirements relevant to the construction requirements for warehouses which (will) contain hazardous chemical substances. These requirements cover the warehouse site and specifications, as well as the storage conditions (including material separation/segregation).

The remaining Chapters of Decision No. 210/2001 cover various areas of chemical substance management specific to certain uses or locales. Chapter IV addresses the management of household, hazardous, healthcare, and sludge wastes. Chapter V regulates the protection of the marine and coastal environment. Chapter VI, somewhat uniquely when compared with other international chemical regulations, addresses "Protection of Earth Crust [sic] from Pollution." Chapter VII relates to air pollution requirements, Chapter VIII addresses "Biodiversity Diversification" (a phrase that is, perhaps at least somewhat, redundant), and Chapter IX lays out the "General Provisions," including "Legal Control" and "Reconciliation Rules." Chapter X will be discussed in more detail in the "Occupational Safety and Health Regulations" section, as the bulk of the data contained in Chapter X regulates the work environment.

In 2015, Kuwait's Standards and Metrology Department, under the auspices of the Public Authority for Industry, notified the WTO of a draft technical regulation, concerning cosmetics and personal care products. The draft regulation, "Cosmetic Products – Safety Requirements of Cosmetics and Personal Care Products" (Standard), is Kuwait's national implementation of the GCC Standardization Organization's (GSO) "GSO 01/ DS 1943/2014" of the same title. The draft addressed the safety parameters and general requirements for all cosmetics and personal care products. It further specified the definitions of the products, as well as the safety requirements, labeling, product claims, packaging, and rules for acceptance and rejection, with a proposed date of entry into force six months from the regulation's publication in the country's Official Gazette.[467]

The scope of the Standard is laid out in Section 1.1 as being applicable to "the safety parameters and general requirements for all cosmetics and personal care products," with an illustrative list of cosmetics being provided in Annex 1 of the Standard.[468] Specifically, the Standard sets out six functions which, if a substance or preparation meets at least one, it may be considered a "cosmetic product": "to clean," "to perfume," "to change the appearance," "to protect," "to keep in good condition," or "to correct body odors."[469] Interestingly, however, Section 1.2 only identifies that these six items may meet the definition – it does not specify where functions must appear descriptively (e.g. on the product label, in marketing material).

Section 3.1 defines "cosmetic and personal care products" collectively as:

> Any substance or mixture intended for use on external parts of the human body (epidermis, hair system, nails, lips and external genital organs) or with the teeth and the mucous membranes of the oral cavity with a view exclusively or mainly to cleaning them, perfuming them, changing their appearance, protecting them, keeping them in good condition or correcting body odors.[470]

Section 3.2 defines a "substance" under the Standard fairly broadly as "a chemical element and its compounds in the natural state or obtained by any manufacturing process, including any additive necessary to preserve its stability and any impurity deriving from the process used but excluding any solvent which may be separated without affecting the stability of the substance or changing its composition," and Section 3.3 defines a "mixture" somewhat reflexively as "a mixture or solution composed of two or more substances."[471] The balance of Section 3 discusses a variety of specific product classifications (e.g. "Sunscreen Product" (Section 3.4), "Fragrance" (Section 3.5), "Colorant" (Section 3.8), and, notably for its inclusion, "Nanomaterial" (Section 3.6), definitions of "Manufacturer" (Section 3.10), "Importer" (Section 3.12) – there is no Section 3.11 in the Standard – and "Distributor" (Section 3.13), and a variety of product packaging-related terms such as "Single application products" (Section 3.15), "Impracticable packaging products" (Section 3.17), and "Non-pre-packaged products" (Section 3.18)).[472]

Section 4 of the Standard, "General Safety Requirements," defines the requirements which cosmetic and personal care products "shall fulfill." Foremost among these, speaking to the Muslim faith, are the requirement that "products shall be completely free from any ingredients that are not aligned with halal practice and rules e.g. Pork, lard [sic]."[473] Further, such products "shall be safe for human health when used under normal or reasonably foreseeable conditions of use," and "homogenous, stable and their properties shall not change during its shelf life and used per the instructions."[474] Section 4.4 provides a list of prohibited substances, primarily those listed in Annex II, as well as several lists of restricted substances, with reference to Annex III, Annex IV (colorants, except for hair-coloring products), Annex V (preservatives), and Annex VI (UV filters). Section 4.6 relates to the requirement to comply with Good Manufacturing Practices (GMP); specifically, "Compliance with good manufacturing practice shall be presumed where the manufacture is in accordance with the relevant harmonized standards such as GSO ISO 22716."[475] As written in the legislation, GSO ISO 22716 is not the only external standard that may be applied to achieve GMPs, but it is specifically enumerated as one in which GMPs will be presumed compliant when followed.

Section 4.8 of the Standard begins a series of sections which address specific criteria for a variety of product types. Section 4.8 itself identifies the microbiological limits for the presence of specific microorganisms. The Section also splits the limits into two categories – "Products specifically intended for children under three years of age, the eye area or the mucous membranes" and "Other products."[476] The microorganisms cited are relatively "common" for such limitations, including *Escherichia coli*, *Pseudomonas aeruginosa*, and *Staphyloccocus aureus*. Section 4.9 sets forth the limitations for substances present in the product, again differentiating between two categories. The first category addresses "the non-intended presence of small quantity [sic] of a prohibited substance, stemming from impurities of natural or synthetic ingredients, the manufacturing process, storage, migration from packaging, which is technically unavoidable in good manufacturing practice…" and states that the foregoing "shall be permitted provided that such presence is in conformity with Article 4.2."[477] Interestingly, the term "small quantity" is not defined either in Section 4.9 or in the Standard as a whole, however. Five heavy metals – lead (10 ppm), arsenic (3 ppm), cadmium (3 ppm), mercury (3 ppm), and antimony (5 ppm) – should not exceed these limits and, in the event they do, "[p]roducts with values above these limits may undergo an assessment to determine the level of risk posed by the product, which would then determine the appropriate enforcement action according to the National Standardization Body (NSB)."[478]

Section 5 relates to the "General Marking and Labeling Requirements" applicable to cosmetic and personal care products, while Section 6 relates to the "Specific Marking and Labeling Requirements." Categories which are listed in Section 6 (e.g. "toilet soap") have more detailed requirements than those in Section 5. Section 5 marking and labeling should include the following information in "indelible, easily legible and visible lettering." All requirements other than the product function and/or use, the warnings and precautionary information, and storage instructions for safe use, which must be presented in both Arabic and English, may appear in Arabic and/or English:[479]

1. The name of the product and the name of [the] trade mark;
2. The name and address of [the] manufacturer or supplier or distributor. The name and the address must be sufficient to identify the undertaking;
3. The country of origin of the product;

4. The nominal content of the product. The nominal quantities shall be expressed in units of weight or volume. The nominal content information is exempted in the following cases:
   a. For products containing less than 5 ml or 5 g
   b. Single-application products
   c. Free samples
5. The date which the cosmetic product, stored under appropriate conditions, will continue to fulfil its initial function and, in particular, will remain in conformity with Article 4.2 [essentially the shelf life];
6. The conditions of use [e.g. warning statements and precautionary statements];
7. Batch number or lot code;
8. Product function [unless such may be "spontaneously and obviously detected"]; and
9. A list of ingredients. [*Note:* these may be printed on the packaging only. For the purposes of this Article of the Standard, an "ingredient" is "any substance or mixture intentionally used in the cosmetic product during the process of manufacturing" except for impurities in the raw materials used and/or "subsidiary technical materials used in the preparation of the cosmetic product but not present in the final product."]

> Section 7 details the claims which may not be included on cosmetic and personal care products; primarily, "pictures and illustrations that are inconsistent with the prevailing social customs and values of the GCC countries," while Section 8 enumerates the packaging requirements for the goods, and Section 9 mandates that "Each consignment of cosmetics and personal care products must be accompanied with a certificate issued either from [the] National Standardization Body (NSB) or from any other official body approved by [the] NSB…"[480]

The balance of the Standard consists of numerous appendices containing the variety of lists discussed in the legislative text of the Standard:

- Annex I: Illustrative List of Cosmetics;
- Annex II: List of Substances Prohibited in Cosmetic Products [approximately 1,328];
- Annex III: List of Substances Which Cosmetic Products Must Not Contain, Except Subject to the Restrictions Laid Down [approximately in excess of 200];
- Annex IV: List of Colorants Allowed in Cosmetic Products [approximately 153];
- Annex V: List of Preservatives Allowed in Cosmetic Products [approximately 57];
- Annex VI: List of UV Filters Allowed in Cosmetic Products [approximately 28]; and
- Annex VII: Symbols Used in Packing/Container.

Pesticide Regulations

As discussed earlier, with respect to the Pesticides Act, Kuwait has transposed it into national law, via "Law No. 21 of 2009 Approving the Pesticides Act in the countries of the Cooperation Council for the Arab States of the Gulf."[481]

On March 30, 2010, Kuwait's Public Authority for Agriculture Affairs and Fish Resources (PAAAFR) announced the implementation of a law on pesticide registration and circulation.[482] The law prohibited the import or manufacture of pesticides without prior permission from the PAAAFR.

For an agriculture or public health pesticide to be registered for import into and use in Kuwait, the applicant company must have a license to import pesticides, or have applied to obtain one from the Ministry of Health.[483] "High risk or extremely poisonous pesticides may not be registered unless proved that other alternatives which are less poisonous are not available. This is decided by the Joint Permanent Committee for the Organization of the Pesticide Manufacturing, Importation and Usage."[484] Five samples of the finished pesticide product and one standard sample of the active substance(s) must be submitted. These samples will be subject to biological experiments and examination to determine their effectiveness on the targeted insect or pest, as well as to chemical analysis.[485]

The label provided must be written in "clear" Arabic and English, and cannot be removed or changed. The label must include the following items:[486]

1. The commercial name of the pesticide;
2. The name of the active substance, its concentration and the ingredients;
3. A copy of the "pesticide equipment" [interestingly, this term is not defined];
4. The name of the producing company and its address;
5. The types of pests affected by the pesticide [target organisms];
6. The proportions of mixture and use;
7. Precautions to be taken in order to protect human beings and untargeted creatures;
8. First aid;
9. Date of pesticide production and date of expiration;
10. Warning signs for safety and security [hazard symbols];
11. A flammable/inflammable sign [hazard symbol];
12. The phrase "Keep the pesticide out of the reach of children";
13. A "clear and remarkable" sign on the containers that contain concentrated and extremely poisonous pesticides;
14. Classification according to the degree of poisonousness;
15. Methods to dispose of the empty containers;
16. Batch number;
17. The size and weight of the container, provided that the weight of the container is not less than (1/4) kg and not more than (1) kg for solid substances and/or not less than (1/4) l and not more than (1) l for liquids for sale in the local market; and
18. [Any applicable] waiting period, in case the pesticide is used against animals or crops.

The list of documentation required to support the registration application is similarly expansive:[487]

1. An original copy of the certificate of manufacturing origin of the pesticide indicating the names of pesticides required to be registered and authenticated by the Chamber of Commerce, the Ministry of Foreign Affairs, and the Embassy of Kuwait in the country of origin;
2. An original copy of the certificate of pesticide usage indicating the names of pesticides required to be registered and authenticated by the Chamber of Commerce, the Ministry of Foreign Affairs, and the Embassy of Kuwait in the country of origin;
3. An original copy of the certificate of agency from the producing company indicating the names of pesticides required to be registered and authenticated by the Chamber

of Commerce, the Ministry of Foreign Affairs, and the Embassy of Kuwait in the country of origin;

4. An original copy of the chemical analysis certificate of the pesticides required to be registered and approved by the producing company;
5. A document showing the method of analyzing the pesticides required to be registered by the producing company;
6. A document that includes the scientific and technical data on the pesticides required to be registered;
7. The original labels on the containers of the pesticides required to be registered must be written in clear Arabic and English and cannot be removed or changed. The labels cannot be accepted if folded;
8. The registration file of the pesticide that includes all the original documents of the pesticide required to be registered and two (2) copies; and
9. A pledge letter from the producing company to reship the pesticides banned from being used or imported pursuant to a decision by the Joint Permanent Committee for the Organization of the Pesticide Manufacturing, Importation and Usage or those that are not consistent with the conditions of the executive regulations.

The fee for first-time pesticide registration, which is valid for a three-year period, is 250 Kuwaiti Dinars.[488] The fee for re-registration every three years, to be paid six months before the expiration of the current registration, is 100 Kuwait Dinars.[489]

Occupational Safety and Health Regulations

There is generally no encompassing OSH-specific legislation in place in Kuwait; however, many companies in Kuwait are adopting international standards. Of specific interest in Kuwait is the Occupational Safety and Health Assessment Series (OHSAS) 18001 standard. The Ministry of Social Affairs and Labor, the Ministry of Health Regulations, and the EPA share authority for management of OSH standards in Kuwait. In addition, the Kuwait Petroleum Council (KPC) maintains its own health, safety, and environmental standards specific to the oil and gas sector, which are in line with international standards.[490]

OSH is primarily managed under a variety of sector- or topic-specific regulations, many of which have been in place for 40+ years. Under such regulations, employers generally have a duty to provide a safe system of work (work processes), a safe work environment (overall facility), and information, training, and supervision on safety and health topics.[491] Relatedly, employees have protection against unfair dismissal (e.g. for being ill) and the ability to receive compensation for unfair dismissal – with no burden of proof on the employee.[492] Examples of such sector- or topic-specific regulations are:[493]

- Law No. 6/2010 concerning Labor in the Private Sector;
- Ministerial Decree No. 164/2006 amending several provisions from Ministerial Decree No. 114/1996 regarding the Requirements and Conditions that Should be Provided in Work Places to Protect the Employees from Work Hazards;
- Ministerial Order No. 56: Protection from Machinery;

- Ministerial Order No. 42, to establish a standing committee to coordinate the work of the Ministry of Social Affairs and Labor and the Ministry of Public Health [sic] in connection with occupational safety and health; and
- Ministerial Decree No. 22, respecting the safety precautions to be taken against occupational injury and disease.

The Private Sector Labor Law (No. 38 of 1964) encompasses any "laborer"; that is, any employee performing manual or mental work in consideration for a wage, provided that they are performing such in the private sector, with some exceptions. Article 40 of the Private Sector Labor Law addresses the protection of workers against accidents, Article 42 relates to requirements for cleanliness, ventilation, lighting, and drainage, and Article 44 mandates the presence of one first aid kit for every 100 workers, and one trained nurse in charge of each kit.[494] With respect to occupational injuries, Article 64 requires employers to pay the full wages of an injured employee for the first six months after their injury, and then 50% of the same thereafter.[495] Article 66 of the Private Sector Labor Law provides a list of occupational diseases, as well as their "causal industries," and Article 68 specifically notes that employers have a shared responsibility in workplace safety, with the employer liability in proportion to the period of employment.[496]

As noted above, Appendix 10 of Decision No. 210/2001 has aspects which relate to safety and health in the workplace. While not exclusive to this environment, Appendix 10-1 presents a definition of "hazardous chemical substances" which tends to mirror the UN's Recommendations. Readers should pay particular attention, however, as there is not a true one-to-one correspondence between Appendix 10-1 and the UN Recommendations. "Hazardous chemical substances" are "chemical materials in their gaseous, solid and fluid states as set under the following classifications":[497]

- Explosives;
- Compressed, liquidated gases or gases dissolved under pressure;
- Inflammable liquids;
- Solid inflammable materials and materials exposed to automatic ignition and materials which, on contact with water, emit inflammable materials;
- Oxidizing factors and organic peroxides;
- Poisonous and contagious materials;
- Radioactive materials;
- Corroding materials; and
- Other dangerous materials.

Additionally, Appendix 3 of Decision No. 210/2001 ("Maximum Limits Allowance for Occupational Exposure to Chemical Substances") lays out the various thresholds under which chemical substance exposure is considered to be of concern.

Waste Regulations

According to a 2014 study by Alsulaili et al., "Currently, there are sixteen landfills in Kuwait. Thirteen are closed; only three are active. Unfortunately, there is not one landfill that meets the criteria of a sanitary landfill. Instead, all of the waste is dumped into random holes."[498] Waste in Kuwait is managed primarily through a pair of Laws which have been examined in other contexts previously: Decision No. 210/2001 Pertaining to the Executive By-Law of the Law of Environment Public Authority and Law No. 42 of 2014 Promulgating the Environment Protection Law.

Safety Data Sheets and Labels

Kuwait does not have any specific regulation(s) in place which mandate the provision and/or use of SDSs.[499] However, importers/exporters of chemical substances must submit the following information to the General Authority for the Environment (GAE):[500]

- A detailed list of the make-up of a chemical product;
- The CAS number of the chemical;
- The environmental impact and its effects on health;
- The purpose for its import or export, and the name of the importer/exporter and the receiving party;
- Safety precautions that must be taken in an emergency;
- The physical and chemical characteristics of the compound;
- The statistical or product serial number, in accordance with UN rules; and
- The "ideal" method for disposal and the container assurance of compliance with the PIC [Prior Informed Consent] and any other international convention to which Kuwait is a signatory.

It appears, anecdotally, that Kuwait will accept SDSs which conform to Regulation (EC) No. 1907/2006 (REACH), Annex II, as amended by Regulation (EU) No. 453/2010.

Similarly, its chemical substance labeling regulations are not clearly delineated. "Chemicals imported into Kuwait must be labeled with the product name, application, active ingredients, percentages of components and composition, United Nations' CAS-NO, side effects, storage/handling/hazmat instructions, environmental and occupational safety health risk, poison control, and disposal instructions."[501]

6

Oman

National Overview

The Sultanate of Oman (Oman) is a country in Southwest Asia, on the southeast coast of the Arabian Peninsula. It borders the United Arab Emirates in the northwest, Saudi Arabia in the west, and Yemen in the southwest. The coast is formed by the Arabian Sea in the south and east, and the Gulf of Oman in the northeast. The country also contains Madha, an enclave enclosed by the United Arab Emirates, and Musandam, an exclave also separated by Emirati territory.[502]

Oman encompasses 309,500 square kilometers of land, an area approximately twice the size of the State of Georgia in the United States.[503] The official language is Arabic, but English, Baluchi, Urdu, and Indian dialects are also spoken.[504]

Governmental Structure

Oman is an absolute monarchy, with 11 governorates or *muhafazah*: Ad Dakhiliyah, Al Buraymi, Al Wusta, Az Zahirah, Janub al Batinah (Al Batinah South), Janub ash Sharqiyah (Ash Sharqiyah South), Masqat (Muscat), Musandam, Shamal al Batinah (Al Batinah North), Shamal ash Sharqiyah (Ash Sharqiyah North), and Zufar (Dhofar).[505] The country employs a mixed legal system of both Anglo-Saxon and Islamic law.

Similar to many of the countries which have been examined thus far, Oman has three branches of government: Executive, Legislative, and Judicial, as follows:[506]

1. Executive
   The head of state is the Sultan, who is also the Prime Minister. He is both chief of state and head of the government. The Sultan holds responsibility for appointing the cabinet members.
2. Legislative
   The legislative branch of the Omani government is bicameral. The Council of Oman is comprised of the Council of State (*Majlis Oman*) and the Consultative Council (*Majlis al-Shura*). The Council of State has 85 seats, and its members are appointed by the Sultan from among former government officials and prominent educators, businessmen, and citizens. The Consultative Council also has 85 seats, but its members are directly elected via a popular vote, with the elected serving renewable four-year terms. Legislation from the Consultative Council is submitted

to the Council of State for review by the Royal Court.[507] Organized political parties are formally banned in Oman.[508]

3. Judicial
   The judicial branch is capped by the Supreme Court, and consists of five judges who are appointed for life. They are nominated by the nine-member Supreme Judicial Council, chaired by the Sultan.

Oman publishes its legislation and related items in its Official Gazette, *Al Jareedah Al Rasmeeyah*, on the first working day of every week, under the auspices of the Law of the Official Gazette, issued by Royal Decree No. 84/2011, as long as there is "proper content" for the Gazette to publish.[509] Unfortunately for many, the Official Gazette is only available in print form, but the index of each issue is published in electronic format on the website of the Ministry of Legal Affairs (http://www.mola.gov.om/).

Key Chemical Regulatory Agencies

The Department of Chemicals (Department) is the responsible authority for the handling and management of chemicals, as well as chemical licensing and importation requirements. The Department falls under the Ministry of Regional Municipalities, Environment and Water Resources (MRMEWR). The Department is organized into two committees: one for policies and government issues, and one for technical cooperation.[510] Incongruous to most of the other Middle Eastern countries, Oman has also developed a chemical substance database. Oman participates in many international treaties, including the Basel, Rotterdam, and Stockholm Conventions.[511]

Key Chemical Substance Regulations

The Basic Law of Oman (RD 101-1996), "which may be regarded as the constitution, considers the protection of the environment and prevention of pollution a social principle and responsibility of the State."[512] As such, much of the authority for environmental protection legislation and related activities, often where some degree of chemical substance legislation may live, ultimately derives from this Law. Relatedly, principal framework legislation is Royal Decree No. 114/2001 "Issuing the Law on Conservation of the Environment and Prevention of Pollution" (RD 114/2001). "This law prescribes strict penalties for the release of environmental pollutants and discharge of effluents, both in the land and the maritime territory of Oman."[513] Other laws correlated with environmental protection in Oman are:[514]

1. RD 10-82, Issuing the Law on Conservation of the Environment and Prevention of Pollution;
2. RD 63-85 and RD 71-89, Amendment of Some Provisions of Law on Protecting the Environment and Pollution Control;
3. RD 34-74, Marine Pollution Control Law;
4. MD 8-84, Discharge of Industrial Water into Public Wastewater Networks; and
5. RD 29-2000, Water Wealth Protection Act.

Royal Decree No. 46 (1995), also known as RD 46-95 "Issuing the Law of Handling and Use of Chemicals" (Decree No. 46), published in the Official Gazette on October 1, 1995 and coming into force on that same day, empowers the Department to issue Regulations and Decisions necessary to enforce Decree No. 46. The Department's roles and responsibilities, which are fairly encompassing, are defined in Article 6 as:[515]

1. To implement the Regulations and ministerial decisions issued to enforce the provisions of this law;
2. To conduct tests on chemicals to determine their toxicity and [the] extent of hazard;
3. To issue Permits [sic] to use, manufacture, import, export or handle hazardous chemicals, according to the stated procedures and decisions issued to enforce this law;
4. To issue Permits, for experimental and scientific research purposes, for any hazardous chemical to research centers, scientific and educational institutions and labs;
5. To prepare inventories and registers for chemicals, as well as their users. Collect related information and data, in addition to amending or canceling the register, so as to allow the concerned agencies to examine and inspect them;
6. To classify chemicals according to local and international classification;
7. To liaise and coordinate locally and internationally to exchange data and decisions pertaining to [the] handling of chemicals;
8. To establish a database for chemicals;
9. To provide technical advice to government and private bodies regarding chemicals;
10. To develop guidelines, programs and rules for staff training in the field of chemicals and promote public awareness for the safe use of chemicals;
11. To check, through inspectors, that all legal conditions are fulfilled, chemical registers [are] examined at site[s] and stop violations of the provisions of the Law and the regulations and decisions implementing it;
12. Present periodic report to the Committee on the Department's activities, list of users registered and permitted to handle hazardous chemicals or any other matters to be submitted to the Committee; and
13. Prepare [an] agenda and make arrangements for the committee meetings, follow up [on] the implementation of its decisions and coordinate between the Committee and the other agencies.

Article 1 of Decree No. 46 regulates "any natural or judicial person who has obtained a permit from the Department [of Chemicals] to handle or use hazardous chemicals," and defines "chemicals" (but not specifically "hazardous chemicals") as "any substance, enlisted, as a hazardous material according to the International classification of hazardous material, which affects the public health and the environment... [except for] explosives [which] are excluded."[516]

As alluded to earlier, under Article 2 of Decree No. 46, "manufacture, import, export, transport, storage, handling and use of any chemical shall comply with the provisions of this Law..."[517] Article 3 of Decree No. 46 establishes the Permanent Committee for Chemicals (Permanent Committee), chaired by the Undersecretary for Environment Affairs.

As per Article 4 of Decree No. 46, the committee established in Article 3 shall have the following responsibilities:[518]

1. Draft Regulations and Decisions required to enforce the Law in accordance with the Sultanate's prevailing enactments and those internationally applied, and follow up their implementation;

2. Develop the procedures and conditions of manufacturing, importing, exporting, transporting, storing, handling and use of chemical [sic], as well as the disposal of their waste, in coordination with the concerned agencies; and

3. Based on [the] recommendation of the Department, investigate and suspend the user from continuing his activities in manufacturing, importing, exporting transporting, storing, handling or using any chemical to avoid any hazard threatening public health and the environment.

With respect to item #3, the Department must notify the user and the concerned agency of a decision to suspend activities, as well as the reasons for which such a decision was taken. The user may submit an appeal to the Minister challenging the decision within 15 days from the date he/she receives written notification of the alleged violation; however, filing an appeal will not result in a suspension of the decision.[519] The Minister will take a decision on the appeal in not more than 30 days from the date the appeal was filed, and there is no further appeal of the decision.[520]

While the ensuing Articles are largely administrative, Article 9 specifically delineates that it "is not permissible to import, export, transport, store or handle any hazardous chemical unless packed in special containers according to the approved and recognized specifications in the Sultanate," and Article 11 requires that the "user of hazardous chemicals shall be committed to dispose of hazardous chemical waste empty containers and any substance in violation of the Law, at his expense and under the supervision of the Ministry, as per the Regulations in force."[521] Article 12 provides a specific "tie in" to OSH requirements, noting that the user of hazardous chemicals must take necessary precautions to protect the workers against health hazards and risks related to their work, and provide them with necessary PPE. Further, the workers must be fully aware of and trained in the "best possible means" of handling and confronting the risks of chemicals. Finally, the employer must prepare a register of the "names of persons, chemicals, quantities, numbers, extent of hazard and methods of handling," which shall be submitted to the Department upon request.[522]

Ministerial Decree No. 248/1997, also known as MD 248-97, "Registration of Hazardous Chemical Substances and the Relevant Permits" (Regulation), addresses the topics specific to its name. The Regulation, which came into force on July 6, 1997, mandates in Article 2 that "Any natural or juridical person who intends to deal with any hazardous chemical by manufacture, import, export, transport, storage, handling, use or disposal shall apply to the Ministry" and obtain a permit.[523] The terminology used in the Regulation is identical to that of Decree No. 46, unless otherwise specified, according to Article 1. The terms which differ are "manufacture," "handling," and "chemical safety data."[524] The Department is tasked with maintaining the list of "hazardous chemicals," according to international classifications. The permit must be maintained, along with the chemical safety data set forth in Annex 2, "in a safe place far from where the chemical is kept or transported," although the term "far from" is not defined in the Regulation.[525]

The classification of chemicals which have "hazardous characteristics" is managed under Annex 1 of the Decree, which explicitly "corresponds to the hazard classification system included in the United Nations Recommendations on the Transport of Dangerous Goods (Revision 5)."[526] The UN Recommendations are contained in the UN Model Regulations, themselves developed by the Committee of Experts on the Transport of

Dangerous Goods of the UN ECOSOC. The UN Regulations address the transport of dangerous goods, grouped into categories, by all modes of transport except by bulk tank. Among these categories are flammable liquids, oxidizers, organic peroxides, and corrosives. The reader should note, however, that Annex 1 contains 13 "characteristics" (categories), not the nine which are found within the UN Recommendations (e.g. Characteristic 10, "Liberation of toxic gases in contact with air or water"; Characteristic 13, "Substances, when disposed, yield harmful materials").[527] In addition, the "characteristics" are not necessarily one-to-one equivalents to the UN Recommendations.

Annex 2 of the Decree contains the requirements for the "chemical safety data" which must be maintained along with the permit, as discussed earlier. In this Annex, Omani authorities lay out guidelines for Chemical Data Sheets (CDS), the term used for SDSs in the country. For consistency, this aspect will be addressed in the "Safety Data Sheets and Labels" section, later.

Ministerial Decree No. 25/2009, also known as MD 25-09, "Issuing the Regulations for Organization of Handling and Use of Chemicals" (Decree 25/2009), published on June 10, 2009 and entering into force the next day, addresses the management of items which its title encompasses.[528] Annex 1 of Decree 25/2009 contains a list of substances whose use is restricted, while Annex 2 of the same contains a list of substances whose use is prohibited. Readers should note that Article 2 of what is effectively the Preamble to Decree 25/2009 specifically cancels Ministerial Decision No. 361/2001 "On the Prohibition of Certain Chemicals." For that reason, Ministerial Decision No. 361/2001 has not been included here.

Article 1 of the text of Decree 25/2009 defines several terms (e.g. "chemical," "user," "handling") similar to those which have been addressed and discussed under previous regulations in this chapter. Article 2 of Decree 25/2009 sets out the specific requirements for storage of chemical substances, with a specific eye toward requirements for warehouses and other types of storage buildings. Article 2 dictates the storage locations (e.g. "store chemicals in designated areas away from industrial activities… and a partition of 10 meter width shall be made between the flammable materials and any source of combustion"), the means by which the products should be arranged and presented (e.g. "they shall be stored in an orderly and harmonized manner with labeling of each chemical showing its common name, chemical composition and degree of risk…"), and the physical construction of the storage area (e.g. "the floor of the store shall be lined with impermeable materials, preventing any shock or electrical short, non-slippery and its walls and structures shall be non-flammable"), among other directions.[529] In addition, it exempts certain classifications of substances via Article 7; specifically, pharmaceuticals, medical drugs, and explosives (as identified in Royal Decree No. 82/77).

Unusually, when compared with other chemical substance regulations in Oman, Article 3 of Decree 25/2009 mandates that the user of any chemical substance stated in Annex 1 must submit the "academic qualifications of the technical working team supervising its use" to the Department, and specifically shall agree "not to sell it in the local market."[530] Additionally, Article 5 requires the user to "be committed to prepare" a contingency plan for the chemical, both inside and outside the establishment.[531]

While Article 3 presumably refers to the management of a spill of such chemical(s), the Article is silent on what aspect(s) this directly addresses. Penalties for violations of the provisions of the regulation incorporate the penalties stated in the Law of Handling and Use of Chemicals, under the auspices of Article 8. If a violation continued for more than one month from the date of the offense, the Ministry may halt the offender's activity under Article 4 of the Law of Handling and Use of Chemicals, and may ban his/her handling or use of chemicals until he/she removes the causes and impacts of the violation at his/her own expense.[532]

Pesticide Regulations

Pesticides are regulated under Royal Decree No. 64/2006 (Decree No. 64/2006) "Issuing the Pesticide Law." There are also other related pesticide regulations in the country, such as Ministerial Decision No. 194 of 2007 (Decision No. 194), published in *Al Jareedah Al Rasmeeyah* No. 854, January 1, 2008, and Ministerial Decree No. 41/2012 (Decree No. 41/2012).

Decree No. 64/2006, available only in Arabic at present, consists of 14 Articles regulating the production, import operations, and trading of pesticides in Oman.[533] Of key importance in Decree No. 64/2006 is the establishment of the responsible authority for the registration and authorization of pesticides that may be imported, exported, manufactured, managed, traded, or used in the country.[534] The Ministry of Agriculture and Fisheries holds this registration and authorization authority under Decree No. 64/2006, as well as for regulating those which are banned or restricted.[535] It further administers the rules for the cancellation of registration(s), sets forth inspection procedures and conditions, identifies the percentage of pesticide residues permissible in agricultural products, and the conditions and procedures for the disposal of pesticides.[536]

Decree No. 41/2012, comprised of 64 Articles, divided into nine Chapters and eight Annexes, entered into force on December 20, 2012, repealing Ministerial Decision No. 194 of 2007, "Restrictions of the List of Forbidden Pesticides and Restricted Use Pesticides." Decree No. 41/2012 is the implementing legislation for Royal Decree No. 64/2006, issuing the Pesticide Law.[537] The reader should not be confused by the timeline – the implementing regulation did in fact occur some six years *after* the June 25, 2006 publication of the Pesticides Law.[538] Decree No. 41/2012 contains general provisions on the entry, registration, manufacturing, circulation, and use of restricted pesticides; analysis and transport of any kind of pesticides.[539] Some key sections of Decree No. 41/2012 are:[540]

- Establishment of the Pesticides Registration Committee (Chapter 1);
- Setting the specific provisions on pesticides registration (Chapter 2);
- Setting the specifications for packaging and labeling (Chapter 3);
- Establishing rules regarding authorization, registration, and procedures for the import and export of pesticides (Chapter 4);
- Regulations concerning movement of pesticides in the State (Chapter 5);
- Regulations regarding manufacture and composition (Chapter 6);
- Establishment of conditions of use of pesticides and allowable limits of residues (Chapter 7);
- Regulations relating to the disposal and also to communication regarding pesticides (Chapter 8); and
- Inspections (Chapter 9).

Decree No. 41/2012 also contains eight Annexes:[541]

1. Fees;
2. Classification of pesticide toxicity according to WHO classification;
3. Terms and conditions of storage by the importers and exporters;
4. The data required to obtain import and export permits;
5. The requirements for storehouses and "exhibitions" of pesticides;
6. Data and documents required to obtain a permit to import the active substance(s);
7. Data and documents required to obtain a license for manufacturing the pesticides; and
8. Disposal requirements for empty containers.

Occupational Safety and Health Regulations

In 2008, the Ministry of Manpower issued Ministerial Decision No. 286/2008, the "Regulation of Occupational Safety and Health for Establishments Governed by the Labor Law" (Regulation). The Regulation, consisting of 43 Articles ranging from lighting to ventilation to noise to occupational diseases, "provides for a comprehensive regulatory framework with the aim of improving safety and health standards in the workplace and protecting workers from various occupational hazards."[542]

In general, the workplace must "support good health by promoting healthy food and physical activity in the workplace, prohibiting smoking in the workplace, and enhancing psychological health and social integration of workers."[543] Specifically, according to Article 15, the employer must take all necessary actions to provide "adequate protection" for the workers' safety while at the workplace.[544] This protection involves the work uniform and PPE, which must comply with safety standards applicable to the particular hazards to which the employees are exposed. It is incumbent upon the employer to train his/her employees on the proper and best ways to use, maintain, and store these items, and to clearly indicate through signage in relevant language(s) where entry is prohibited without the use of PPE.[545]

Furthermore, the Regulation provides a specific list of measures which employers must implement, to minimize occupational accidents as well as the exposure to various risks and hazards including: fire, mechanical and electrical risks, chemical hazards, heavy duty machinery, workers' transport vehicles, and other items.[546] The Regulation is highly granular in its level of detail – examples of which can be seen with respect to:

- Lighting in the workplace:[547]
  The employer must provide sufficient, adequate, natural or artificial lighting, distributed in the workplace equally, free from direct or reflective rays, in addition to a system of emergency lighting in case of the failure of the normal lighting. The lighting system must clearly show emergency exits so that the workers can locate and use them.
- Air quality:[548]
  Polluted air shall be avoided by providing a natural or artificial ventilation system that provides fresh air in the workplace and uses local ventilation where sources of pollution exist. This system must effectively suck the polluted air out. The Regulation

also specifies the minimum percentage of oxygen, speed of air, and the maximum degree of relative humidity in the workplace.

- Noise:[549]

To protect the workers from exposure to noise, noisy operations that exceed the permissible levels must be isolated away from the workers, or sound-insulated rooms should be used. Additionally, insulating, absorbing, or reflective equipment should be installed on noisy machines.

Uniquely, the Regulation also addresses the specific OSH needs related to women, as well as to people with special needs. Employers, for instance, must not expose women to materials or occupational practices which could adversely impact the safe delivery of children, or the safety and health of the fetus.[550]

Waste Regulations

Ministerial Decision No. 18/93, "Regulations for the Management of Hazardous Waste" (Decision No. 18/93) was published on February 2, 1993 in *Al Jareedah Al Rasmeeyah*, and entered into force on March 1, 1993. It was later amended in 2002 by MD 56-2002, "Amendment of Regulation on MD 18-93" [sic].[551] As per Article 1, Decision No. 18/93 was to be managed under the auspices of the Ministry of Regional Municipalities and the Environment (Ministry).[552] Interestingly, Decision No. 18/93 is mostly a procedural regulation; it mandates permits for a variety of hazardous waste-related activities, but does not include items which are generally found in the same type of regulation for other jurisdictions globally (e.g. how a hazardous waste is classified as such, requirements for segregation, packaging types, on-site storage time limits, and so forth).

Article 1 of Decision No. 18/93 defines "hazardous waste" in a manner somewhat similar to other global regulations on the topic; however, it is simultaneously unusually general in its scope:

> Any waste arising from commercial, industrial, agricultural or any other activities which, due to its nature, composition, quantity or any other reason is: hazardous or potentially hazardous to human health, to plants or animals, to air, soil or water. This includes explosive, radio-active [sic] or flammable substances; [sic] which may cause disease as well as those issued by a decision from the Minister.[553]

Relatedly, Article 1 defines a "hazardous waste generator" as either "the owner (and/or his agent) of any land or premises of any type where hazardous waste is generated" or "any person (and/or his agent) trading in hazardous materials having residues of any kind or from any source."[554] The remaining terms defined in Article 1 address "transporter," "storage facility," and "pretreatment," among others.

Article 2 of Decision No. 18/93 begins the crux of the regulation, dictating that the Ministry shall be responsible for preparing a standard process (format) for issuance of a Hazardous Waste License, which is required to be held by a hazardous waste generator. The applicant for such a license must explain how they will apply the BAT to their waste production and operational practices, in order to minimize the generation of such hazardous waste, including but not limited to "practicable recycling procedures."[555] Further, any holder of a license which was previously issued under Royal Decree No. 114/2001 must submit an application to the Ministry.

Article 5 of Decision No. 18/93 codifies the requirement for a "Consignment Note" to be completed for "each category of hazardous waste before the hazardous waste leaves his land or premises."[556] A "Consignment Note" is essentially a hazardous waste manifest, and is defined in Article 1(7) as "a document listing the category and quantity of hazardous waste in accordance with the relevant order issued by the Minister."[557] Article 8 stipulates that every generator of hazardous waste must store such in "approved storage facilities on his land or at his premises until its removal in accordance with the terms of the license issued by the Ministry."[558] As per Article 9, hazardous waste must be transported by licensed transporters, and such licenses will be issued "with conditions after the approval of Royal Oman Police." However, these "conditions" are not expressly defined in Decision No. 18/93.

The balance of the Articles are largely straightforward with respect to the procedural management and disposal of hazardous waste. Article 10 mandates the application for a license by every owner of any site where hazardous waste is to be stored, and that he/she shall ensure that all hazardous waste received at the site is accompanied by a Consignment Note.[559] Article 11 dictates the same for owners of a storage facility, Article 12 the same for owners of a hazardous waste pretreatment site, and Article 13 the same for the owner of any hazardous waste landfill site.[560] Finally, Article 14 gives the Minister and/or his staff the right to inspect any activities related to the generation of hazardous waste.

Article 15 prohibits the import of hazardous waste into Oman without a permit from the Minister, and Article 16 prohibits the issuance of a hazardous waste license/permit from the Ministry without the express approval of the Ministry of Health in the areas of collection, transport, storage, pretreatment, and disposal of hazardous waste.[561]

Safety Data Sheets and Labels

As noted earlier, CDS is the term used for SDSs in Oman. Ministerial Decree No. 248/1997, "Issuing the Regulation for the Registration of Chemical Substances and the Relevant Permits," which has also been discussed previously, addresses some CDS aspects. CDSs are primarily managed through Annex 2 of Decree No. 46. A CDS contains 13 categories, or sections, plus one for "Other Information":[562]

1. Scientific name;
2. Common name;
3. CAS number;
4. Chemical and physical properties;
5. Composition;
6. Stability and reactivity [data];
7. Toxicity, and "hazard to man and environment";
8. Safety precautions;
9. First aid and accidental release measures;
10. Transport information;
11. Packing, handling, and storage measures;
12. Transport information;
13. Disposal considerations; and
14. Other information (e.g. sample expiry date).

Readers should pay particular attention to the fact that the above 13 categories' data must be certified by the exporter, the manufacturer, the producer, or a laboratory recognized by the Ministry. If the original information received from the source was not complete enough to satisfy this requirement, then additional certified papers may be attached. Finally, such data – or a certified copy thereof – must be kept "near the chemical, for easy reaching and review during handling."[563]

Ministerial Decision No. 317/2001 (Decision 317/2001), "Issuing the Regulations for the Packing, Packaging, and Labeling of Hazardous Chemicals," issued under the auspices of the MRMEWR, lays out the labeling instructions and the specific requirements applicable to containers.[564] It was published in *Al Jareedah Al Rasmeeyah* on December 10, 2001, and came into effect on June 1, 2002.[565] Decision 317/2001 is applicable, under Article 1, to "any person whatsoever, who imports, exports, transports, stores, handles, uses or discharges any hazardous chemical," except those which are specifically excluded in Article 4 [pharmaceuticals and medical drugs, explosives (again, as per Royal Decree 82/77) and radioactive materials].

Article 1 goes on to list fairly specific requirements for such containers, directing that, among other aspects:

- Firstly, concerning containers:
    - Containers shall be strong, vibration proof, and resistant to damage, whether this may be caused either by the external environment or the contained substance;
    - Containers shall be previously unused and shall be leak proof and non-reactive; and
    - Containers shall be firmly sealed so as to prevent any leakage.[566]

Next, Article 1 addresses the requirements for labels on such containers:

- Secondly, concerning labeling:
    Labels shall be firmly fixed to the containers and shall clearly include not less than the following details:
    - Both the scientific and commercial names of the contents plus the quantity;
    - The chemical and physical properties of the contained substance;
    - The degree of risk of the contained substance plus its international hazard classification in both the Arabic and English Languages [sic];
    - Chemical safety guidelines for handling or dealing with chemicals, especially during any incident or emergency;
    - The purpose of use of the contents and the date of expiry;
    - The full name and address of the manufacturer or producer;
    - Storage instructions relating to temperature, pressure, light, and so forth;
    - All labels shall be clearly written, easily readable, and firmly fixed to the container; and
    - Labels and tags shall be damage resistant, non-flammable, and shall not be readily removed.[567]

Article 2 directs that "All such relevant hazardous chemical warning symbols as depicted in the attached Appendix, shall also be drawn on, or fixed to, each container."[568] Interestingly, the Article employs the phrase "drawn on"; however, it is believed that this means (pre)printed on the packaging itself, as opposed to a "sticker type" label, which would likely meet the "fixed to" requirement.

Finally, Article 3 sets out the specific penalty for violations of Decision 317/2001, noting that "any violating activity concerning chemicals, or any violation of the provisions of these Regulations shall cause the offending person to cease to be allowed to practice his activity" – tantamount to a complete cessation of operations.[569]

7

Pakistan

National Overview

Formally known as the Islamic Republic of Pakistan (Pakistan), with Islamabad as its capital, the country was established in 1947 when British India split into the Muslim state of Pakistan (with West and East sections) and largely Hindu India.[570] In 1971, East Pakistan became the separate nation of Bangladesh.[571] In 2013, for the first time, Pakistan had a democratically elected government complete a full term, and it transitioned to a successive, democratically elected government.[572]

Pakistan is comprised of four provinces, one territory, and one capital territory. "Each Province is headed by a Governor and Provincial Cabinet, all of whom are appointed by the chief executive. The Northern Areas and Federally Administered Tribal Areas are administered by the federal government but enjoy considerable autonomy."[573]

It is worth mentioning that the government of Sindh is the government of the province of Sindh, Pakistan. As with several other Middle Eastern countries, at times a region or province of the country takes more stringent action with respect to chemical, environmental, or other aspects in its area of jurisdiction than the country does. This is the case in Sindh, which has gone so far as to form its own Environmental Protection Agency (EPA). While the Sindh EPA and its related regulations will not be formally discussed in this text, as we instead examine Pakistan as a whole, readers should be aware of the existence of the regulatory body, and may wish to become more familiar with it. The website of the Sindh government is at http://www.sindh.gov.pk/, and the site of the Sindh EPA Environment Climate Change & Coastal Development Department may be accessed at http://epasindh.gov.pk/. However, because it is so thoroughly developed, later in the text the "Sindh Environmental Protection Act, 2014," which entered into force on December 16, 2014, will be discussed in some detail.

Governmental Structure

Pakistan is a constitutionally based parliamentary democracy that became independent in 1947. The official language is Urdu.[574] The initial Constitution, which was created in 1973 and lays out Pakistan's governmental structure, was suspended for three years in October 1999. Under the Constitution, and similar to other countries in the region, the

Chemical Regulation in the Middle East, First Edition. Michael S. Wenk.
© 2018 John Wiley & Sons Ltd. Published 2018 by John Wiley & Sons Ltd.

President is the head of state and the Prime Minister is the head of the government. The Constitution mandates that both leaders be Muslims.[575]

> The President, who must be a member of the National Assembly, is elected to a five-year term by an electoral college consisting of Members of both houses of Parliament and Members of the Provincial Assemblies. The Prime Minister is selected by the National Assembly and serves a four-year term.[576]

"Under the Constitution, the government of Pakistan is obligated to bring all laws into conformity with Islam. To achieve this objective, many statutes based on Islamic injunctions have been enacted."[577]

The Pakistani legislative branch is comprised of a bicameral Parliament known as *Majlis-e-Shoora* ("Councils"). "The *Majlis-e-Shoora* consists of the Senate, whose 104 Members are indirectly elected by the Provincial Assemblies to six-year terms, with one half of the membership renewed every three years, and the National Assembly," which is comprised of 342 members.[578] Of the 342 National Assembly members, 60 must be women and 10 must be minorities (non-Muslims).[579] All National Assembly members are directly elected by the people to five-year terms.[580]

The judicial branch of the Pakistani government is comprised of the Supreme Court, Provincial High Courts, a Federal Islamic (or *Shari'a*) Court, and various provincial and district civil and criminal courts.[581] Pakistan employs a common law system with Islamic law influence.[582] With respect to the Supreme Court, Pakistan's highest court, justices are nominated by an eight-member parliamentary committee, which itself receives nominations from the Judicial Commission.[583] The Chief Justice of the Supreme Court is complemented by 16 other judges.[584] "Each Province has a High Court, the justices of which are appointed by the president after conferring with the Chief Justice of the Supreme Court and the Provincial Chief Justice."[585]

Governmental legislation is published in the Gazette of Pakistan, which has two issues: weekly and extra-ordinary.[586] The weekly Gazette consists of two parts: Part I contains notifications regarding appointments, promotions, and so forth issued by various Ministries or Divisions of the government and/or the Supreme Court of Pakistan, aside from the Ministry of Defense;[587] Part II contains notifications issued by the Ministry of Defense. The Gazette may be accessed at https://web.archive.org/web/20051214073450/http://www.pakistan.gov.pk/gazpak.jsp.

Key Chemical Regulatory Agencies

> Pakistan lacks the expert capacity for data collection, analysis and dissemination to various stakeholders… The chemical industries do not maintain safety and security related data in a transparent manner. No data exists regarding areas heavily polluted by chemical waste and sites of obsolete chemicals. Moreover, [a] national information system regarding chemicals and related safety and security aspects do [sic] not exists [sic].[588]

Within Pakistan, regulations specific to the management of chemical substances are administered by multiple agencies: the Ministry of Environment, the EPA, and the Ministry of Industries and Production. The EPA, also known as "PAK-EPA," was established under Section 5 of the Pakistan Environmental Protection Act (PEPA) of 1997, which came into force on December 3, 1997. Readers should pay careful attention to the similarity in names between these two agencies – PAK-EPA is the competent authority, and PEPA is the legislation. Further, as noted in the opening section of this chapter, readers should recall that there is PAK-EPA, as well as the Sindh EPA. Each mention of the respective Agencies herein will be clear with respect to its province.

As per Section 5, the "Federal Agency," as it is known in PEPA, shall be headed by a Director General who will be appointed by the Pakistani government. Section 6 of PEPA extensively defines the specific functions of PAK-EPA. Several of the key items are:[589]

1. Administer and implement the provisions of this Act and the rules and regulations made thereunder;
2. Prepare, in coordination with the appropriate Government Agency and in consultation with the concerned sectoral Advisory Committees, national environmental policies for approval by the Council;
3. Take all necessary measures for the implementation of the national environmental policies approved by the Council;
4. Prepare and publish an annual National Environment Report on the state of the environment;
5. Ensure enforcement of the National Environmental Quality Standards;
6. Establish standards for the quality of the ambient air, water, and land, by notification in the Official Gazette, in consultation with the Provincial Agency concerned;
7. Coordinate environmental policies and programs nationally and internationally;
8. Establish systems and procedures for surveys, surveillance, monitoring, measurement, examination, investigation, research, inspection, and audit to prevent and control pollution, and to estimate the costs of cleaning up pollution and rehabilitating the environment in various sectors;
9. Certify one or more laboratories as approved laboratories for conducing [sic] tests and analysis and one or more research institutes as environmental research institutes for conducting research and investigation, for the purposes of this Act;
10. Identify the needs for, and initiate legislation in various sectors of the environment;
11. Render advice and assistance in environmental matters; and
12. Specify safeguards for the prevention of accidents and disasters which may cause pollution, collaborate with the concerned person in the preparation of contingency plans for control of such accidents and disasters, and coordinate implementation of such plans.

Also under PEPA, this time in Section 3, the Federal Government established the Pakistan Environmental Protection Council. Council members hold office for a term of three years, and comprise:[590]

1. Prime Minister or such other person as the Prime Minister may nominate in this behalf (Chairperson);
2. Minister Incharge [sic] of the Ministry or Division dealing with the subject of environment (Vice Chairperson);
3. Chief Ministers of the Provinces;
4. Ministers Incharge [sic] of the subject of environment in the provinces;

5. Such other persons not exceeding 35 as the Federal Members Government may appoint, of which at least 20 shall be non-official including five representatives of the Chambers of Commerce and Industry and Industrial Associations and one or more representatives of the Chambers of Agriculture, the medical and legal professions, trade unions, and non-governmental organizations concerned with the environment and development, and scientists, technical experts and educationists; and
6. Secretary to the Government of Pakistan, in-charge [sic] of the Ministry or Division dealing with the subject of environment (Secretary).

The Environmental Protection Council has substantially smaller functions and powers than PAK-EPA, specifically:[591]

1. Coordinate and supervise enforcement of the provisions of this Act;
2. Approve comprehensive national environmental policies and ensure their implementation within the framework of a national conservation strategy as may be approved by the Federal Government from time to time;
3. Approve the National Environmental Quality Standards;
4. Provide guidelines for the protection and conservation of species, habitats, and biodiversity in general, and for the conservation of renewable and non-renewable resources;
5. Coordinate integration of the principles and concerns of sustainable development into national development plans and policies; and
6. Consider the National Environment Report and give appropriate directions thereon.

Key Chemical Substance Regulations

While relevant Ministries and organizations maintain their data with respect to the chemical substances under their purview, no central database of such exists. Indeed, "[n]o data exists regarding areas heavily polluted by chemical waste and sites of obsolete chemicals. Moreover, national information system regarding chemicals and related safety and security aspects do not exists [sic]."[592]

According to Jaspal and Haider (2014), "there are 53 Acts and regulations, which are considered relevant for the management of chemicals in Pakistan. However, in reality, most of these acts have just some sort of relevance with chemicals."[593] The statutes which do contribute substantively to the management of chemicals are few in number, as will be discussed after an examination of the Sindh government's Hazardous Substances Rules (2014).

As discussed earlier, within Pakistan, certain provinces have established their own regulations with respect to the management of chemical substances. The government of Sindh, seated in Karachi, and the government of the province of Sindh established the Hazardous Substances Rules (Rules). Originally issued in 2007 and amended on December 16, 2014, the Rules refer to the management of chemical substances in the Sindh province. The Rules are somewhat unique in that they encompass regulations for a multitude of areas, such as labeling, safety and health, and others, where other statutes would normally parse these categories out into other more specific regulations.

Interestingly, "hazardous substances" are not defined expressly in Section 2 of the Rules, the definitions section. Instead, Section 3 declares "substances prescribed as hazardous substances" to be those which are specifically listed in Schedule I. Section 4 mandates that an application for a license shall be made by a person wishing to import or transport a hazardous substance, with the specific information required listed in Form A of the Rules, as well as in Sections 20 and 21.[594] Further, Section 5 requires the completion of an EIA as part of the license application mandated in Section 4. Specifically, "An application… shall be accompanied by an environmental impact assessment of the project or industrial activity involving generation, collection, consignment, transport, treatment, disposal, storage, handling or import of a hazardous substance in respect of which the license is sought."[595] Section 5 goes on to prescribe the content of the EIA, as well as the attendant Safety Plan.

Section 6 discusses the process for review of the application by the Authority, and Section 7 procedurally outlines the process for paying the associated fee and receiving the license. Sections 8(a)–8(g) of the Rules detail the explicit conditions with which the licensee must comply, specific to technical knowledge, worker safety, packaging and labeling, and training to be conducted. Explicitly:[596]

(a) The licensee shall employ qualified technical personnel having [the] necessary knowledge and experience regarding the use, storage and handling of the hazardous substance, and safety precautions relating thereto;
(b) The hazardous substance shall be packed and labeled in accordance with rule 9;
(c) The premises of the licensee shall comply with the conditions laid down in rule 10;
(d) The licensee shall ensure compliance with the provisions of rules 11 and 12 regarding safety precautions;
(e) The licensee shall provide necessary information, and where required training, to the persons to whom the hazardous substances are sold or delivered, regarding the use, storage and handling of the hazardous substances, and safety precautions relating thereto;
(f) The licensee shall maintain a detailed record of the quantity, type, quality and origin of the hazardous substance and the names and addresses of the persons to whom the hazardous substances are sold or delivered; and
(g) The licensee shall not extend his operation beyond the scope of the project or industrial activity in respect of which the environmental impact assessment has been submitted and approval granted.

The reader should note that the language of the Rules uses the term "rule" instead of "section" throughout. For clarity of understanding relating to this regulation (e.g. potentially causing confusion between the shortened version of the regulation name – Rules and the sections therein – rule), the author uses the term "section" to mean "rule." One may also see in Section 8 an illustration of the earlier point regarding how a variety of regulations normally found in separate statutes are included in the Rules.

Section 9, Packing and Labeling, sets forth the requirements for the proper container type to be used for a "hazardous substance," as well as the items which should appear on the container's label. The container:

> shall be of such size, material and design as to ensure that – (a) it can be stored, transported and used without leakage and safely[, and] (b) the hazardous substance therein does not deteriorate in a manner as to render it more likely to cause, directly or in combination with other substances, an adverse environmental effect.[597]

Additionally, "the following information shall be printed conspicuously, legibly and indelibly on every container of a hazardous substance":[598]

a. Name of the hazardous substance;
b. Name, address and license number of the licensee;
c. Net contents (volume or weight);
d. Date of manufacture and date of expiry, if any;
e. A warning statement comprising –
    i. The word "DANGER!" in red on contrasting background
    ii. A picture of a skull and cross-bones
    iii. Pertinent instructions for use, storage and handling and safety precautions relating thereto
f. Instructions regarding return or disposal of the empty container; and
g. Basic instructions mentioning immediate steps to be taken in case of any accident or emergency, preferably in the local language.

Of particular note in the requirements of Section 10 of the Rules, "Conditions for premises," are the regulations which address "the premises in which a hazardous substance is generated, collected, consigned, treated, disposed of, stored or handled."[599] In addition to complying with the conditions specified in Schedule IV of the Rules (examined later), a "notice" containing the following information must be placed on the "outer door or gate" of the premises: "the words 'DANGER ! [sic] HAZARDOUS SUBSTANCE!' in red, on a contrasting background; and a prominent picture of a skull and cross-bones."[600] If the substance(s) is/are imported, approval from the Climate Change Division shall be obtained.[601] Of note, however, is that the Rules do not distinguish whether the hazardous substance is an Ozone-Depleting Compound (ODC) or other environmentally sensitive substance. Rather, the approval is required for any type of imported hazardous substance.

Sections 11 and 12 of the Rules detail specific requirements relating to "General Safety Precautions" (Section 11), which address the "safety precautions [to be] conveyed to persons to whom the hazardous substances are sold or delivered" and "Safety Precautions for Workers" (Section 12). As noted earlier, the Rules tend to incorporate language which is often addressed in other statutes in a country's regulatory scheme, and these Sections provide two such examples. Section 11, "General Safety Precautions," lays out a list of five specific safety precautions which the licensee must undertake. Among these are: to instruct the persons to "carefully read and follow the instructions and safety precautions printed on the container" (which shall be either in Urdu or the local language), to wear proper PPE, to avoid contact with the hazardous substance via the skin and eyes – as well as clothing, and to not eat, drink, or smoke within the vicinity of the hazardous substance.[602] It should be observed that Section 11 does not comprise an exhaustive list of safety precautions, but serves only as a minimum.

Section 12, "Safety Precautions for Workers," relates precisely to that – workers employed by the licensee for the purpose of handling hazardous substances. This Section is considerably more extensive in its prescriptive content than Section 11. Among its key requirements are: a prohibition on employment of workers younger than 18 or older than 60 for jobs involving the physical handling of hazardous substances, the thorough training of workers and their supervision by "qualified supervisors," a requirement for the provision of PPE, and also a ban on workers who are not wearing such from performing work with hazardous substances. Further, Section 12 requires the checking and maintenance of fire-fighting, emergency, and safety equipment, the establishment of a first aid medical facility on the premises – supervised by trained staff and equipped with required antidotes, medical check-ups of workers both at the time of hire and at least once a year thereafter, and a record of each worker "containing, amongst other details, his name and address, his medical check-up history, and the hazardous substances handled by him."[603]

The licenses mentioned in Section 4 of the Rules are further discussed in Section 13, noting that such licenses are valid for an initial period of one year, and if a renewal application is made for the same under Section 14, the license in force will remain valid until a decision on the renewal is taken. Licensees applying for renewal should, a minimum of 30 days before the expiration date, provide an update to the initial EIA, unless conditions have changed such that a new EIA would be required.[604] Renewal applications should be provided on Form A of Schedule II of the Rules. Bear in mind that if the Authority believes the conditions of the license have not been complied with, or that information provided in the EIA is incorrect, it will issue a "show cause" notice to the licensee. He/she must then provide reason(s), within two weeks of receipt, why the license should not be cancelled.[605]

Under the Rules, the authority to enter and inspect the premises, at least once per year, where the hazardous substances are being generated, collected, consigned, treated, disposed of, stored, or handled, is granted to the EPA.[606] Further, the licensee is obligated to cooperate with duly authorized staff, and to provide any information required.

One such piece of information which the Authority may elect to review is the Safety Plan, mandated under Section 5(2)(a), and whose specific content, enumerated in Section 17, is:[607]

a. An analysis of major accident hazards relating to the hazardous substance involved;
b. An assessment of the nature and scope of the adverse environmental effects likely to be caused by major accidents;
c. A description of the safety equipment and systems installed and safety precautions taken; and
d. A description of the emergency measures proposed to be taken on and off the premises of the applicant to control a major accident, and to mitigate its adverse environmental effect.

It should be noted that the Safety Plan must be reviewed and approved by the Authority, "in consultation with other relevant Government Agencies and Departments including the licensee."[608] The manifest purpose of this cooperative review is to ensure that it covers "all anticipated contingencies and all emergencies likely to result from a major accident involving the hazardous substance involved, and that the concerned Government Agencies, Departments and the licensee are aware of their specific responsibilities thereunder."[609] A requirement of the Rules, in Section 18, and related to the Safety Plan, is the requirement for the licensee to notify the Authority of any "major" accident occurring on their premises, as well as submit a report in the form proscribed in Schedule V, both within the first 24 hours and weekly thereafter.[610]

As examined previously in Section 4, the import of hazardous substances is regulated by Section 20 of the Rules. Such import requires a license, the information set forth in Form A of Schedule II, and the following: the name of the port of entry into the province of Sindh, how the substance will be transported from the exporting country to Pakistan, the quantity of the hazardous substance being imported, complete information with respect to the safety precautions to be adopted, the purpose for which the hazardous substance will be utilized, along with the EIA (if one is required under Section 5), and a copy of approval from the Ministry of Climate Change, International Cooperation (IC) Wing, of the government of Pakistan.[611]

Finally, under Section 21, the entity seeking a license for the transport of hazardous substances must provide, again in addition to Form A of Schedule II, the following data points: the name and address of the person from whom the hazardous substance is to be collected and to whom the substance is to be delivered, the quantity to be transported, and the mode of transport (including the "full particulars and specifications of the motor vehicles or other conveyance"), the route to be used, the date and time proposed for the transport, the nature of the waste (that is, the physical form), its toxicity and a SDS, and a copy of the contingency or emergency response plan.[612] To be sure, this last point is especially challenging to manage, as logistics managers and others may not know the precise route of transport or the means of conveyance (e.g. in the case of Less-Than Truckload (LTL) quantities).

Returning to a deeper examination of the PEPA, alternatively known as Act No. XXXIV of 1997, while the title of the Act gives the appearance of regulating only aspects related to environmental protection, Section 2(xviii) specifically defines a "hazardous substance" whose definition incorporates a variety of physical and chemical properties:[613]

(a) a substance or mixture of substance, other than a pesticide as defined in the Agricultural Pesticide Ordinance, 1971 (II of 1971), which, by reason of its chemical activity is toxic, explosive, flammable, corrosive, radioactive or other characteristics causes, or is likely to cause, directly or in combination with other matters, an adverse environmental effect; and

(b) any substance which may be prescribed as a hazardous substance.

Section 14, "Handling of Hazardous Substances," is perhaps the most directly relevant section of PEPA to chemical substance management. The Section directs that "no person shall generate, collect, consign, transport, treat, dispose of, store, handle or import any hazardous substance," except under a license issued by the "Federal Agency" or in accordance with "the provisions of any other law for the time being in force, or of any international treaty, convention, protocol, code, standard, agreement or other instrument to which Pakistan is a party."[614] Thus, international treaties such as the Basel, Rotterdam, Vienna, and Stockholm Conventions, as well as the Montreal and Kyoto Protocols, are implicitly incorporated.[615] Section 16, "Environmental Protection Order," addresses, in Section 16(1), the measures which the "Federal Agency" may take to address "discharge or emission of any effluent, waste, air pollutant or noise, or the disposal of waste, or the handling of hazardous substances or any other act or omission" which it believes is "likely to occur, or is occurring or has occurred in violation of the provisions of this Act, rules or regulations or of the conditions of a license, and is likely to cause, or is causing or has caused an adverse environmental effect…"[616] Note that Section 16(1) specifically states "is *likely* to occur" and "is *likely* to cause." As such, the "Federal Agency" has the legal authority to take steps to prevent adverse environmental effects, even if such has not yet actually occurred. These steps may involve immediate stoppage of the process(es), requiring installation, etc. of equipment designed to control or prevent the issue, removal of that which has caused/may cause an adverse environmental effect, and/or remediation of the affected site.[617]

Section 17 details the penalty structure for violations of PEPA, which may range from 1,000 Pakistani rupees (PKR) to 1,000,000 PKR, as well as potential imprisonment, civil compensation, confiscation of the means of production, and/or a complete cessation of operations.[618] Section 20 provides for the establishment of "Environmental Tribunals" consisting of a "Chairperson who is, or has been, or is qualified for appointment as, a Judge of the High Court to be appointed after consultation with the Chief Justice of the High Court and two members... appointed by the Federal Government."[619] Of these two members, "at least one shall be a technical member with suitable professional qualifications and experience in the environmental field as may be prescribed."[620] Environmental Tribunals are to be established by the Federal Government, by notification in the Gazette of Pakistan, to adjudicate violations of PEPA.

In January 2016, the Ministry of Climate Change promulgated a Statutory Notification (S.R.O.) entitled the "Handling, Manufacture, Storage, Import of Hazardous Waste and Hazardous Substances Rules, 2016" (Hazardous Substances Rules), which came into force immediately. The definition of "hazardous substance" in Schedule 2(c) of the Hazardous Substances Rules is the same as provided in Section 2(xxv) of PEPA. As its name suggests, the Hazardous Substances Rules encompass all aspects relating to the handling, manufacture, storage, and import of hazardous waste and hazardous substances, but specifically "an industrial activity in which a hazardous chemical, which satisfies any of the criteria laid down in Part I of Schedule 1 or listed in Column 2 of Part II of this Schedule is, or may be, involved" (Schedule 3(1)(a)) and "isolated storage of a hazardous chemical listed in Schedule 2 in a quantity equal to or more than the threshold quantity specified in Column 3, thereof" (Schedule 3(1)(b)).[621]

The remaining Schedules of the Hazardous Substances Rules are voluminous; as such, only the key Schedules will be examined. Schedule 3 addresses the practical roles and responsibilities for the "occupier" (*Note:* the term is not specifically defined) who has "control of an industrial activity" defined in Schedule 3(1)(a) and 3(1)(b). Such an "occupier" must both identify the major accident hazards and take "adequate steps" to prevent such major accidents, limit the consequences of the same to people and the environment, and provide on-site workers with the information, training, and equipment – including "antidotes" – necessary to ensure their safety.[622] As per Schedule 4, when a major accident occurs on a site or in a pipeline, the occupier must, within 48 hours, notify the appropriate agency of the accident, as well as provide PAK-EPA with a report relating to it.[623] PAK-EPA will then undertake a full analysis of the major incident.[624]

Schedule 6 directs that an occupier may only undertake industrial activity after being granted Environmental Approval for the activity, and after submitting a written report detailing the items required in Schedule 7 at least three months before beginning activity. PAK-EPA has 60 days from receipt of the report to take a decision on the activity proposed.[625] Schedule 9 requires the occupier to prepare an environmental audit report on the industrial activity, containing the items specified in Schedule 8. Such a report must be sent to PAK-EPA at least 90 days before beginning the specified activity.[626] Further, as per Schedule 9(4), the occupier of both new and existing industrial activities must have an independent safety and environmental audit of the industrial activities conducted by a certified environmental auditor who is not associated with the activity.[627] The occupier must forward a copy of the report, along with his/her comments on it, within 30 days of the completion of the audit. Finally, this report must be updated annually by "conducting a fresh safety and environment audit and forward[ing] a copy of it with his comments thereon within to days [sic] to the PAK-EPA."[628]

Schedule 12 of the Hazardous Substances Rules mandates the occupier to prepare and keep up-to-date an on-site emergency plan containing the details required in Schedule II. Notably, "… that plan shall include the name of the person who is responsible for safety on the site and the names of those who are authorized to take action in accordance with the plan in case of an emergency."[629] The reader should be aware that the Schedule specifically calls for "the *name* of the *person*" and also the *names* of those authorized [emphasis the author's], and not just a position title.[630] Furthermore, the occupier is required by Schedule 12(4) to conduct a mock drill of the emergency plan every 12 months and Schedule 12(5) requires a "detailed report of the mock drill" to be made "immediately available to the concerned Authority."[631] PAK-EPA also has emergency plan obligations under the Hazardous Substances Rules, via Schedule 13(1). Here, the Authority must "prepare and keep up-to-date an adequate off-site emergency plan containing particulars specified in Schedule 12 and detailing how emergencies relating to a possible major accident on that site will be dealt with…"[632] Additionally, "in preparing that plan the PAK-EPA shall consult the occupier, and such other persons as it may deem necessary."[633]

Schedule 14 directs the occupier to "take appropriate steps to inform persons outside the site either directly or through District emergency units who are likely to be in an area which may be affected by a major accident."[634] As such, Schedule 14 mandates what is a largely unique aspect among Middle Eastern chemical substance regulations, and certainly among Pakistani regulations: direct communication with those potentially affected by an industrial accident on a proactive basis. Furthermore, Schedule 14(2) requires the occupier to inform persons about the industrial activity (not just relating to a major accident) within 90 days of commencement of operations.[635]

Schedule 17 of the Hazardous Substances Rules mandates the provision of the following information at least 30 days, or as reasonably soon as possible but not later than the date of import, prior to the importation of hazardous substances into the country. As per Schedule 17(2), such information should include:[636]

(i) The name and address of the person receiving the consignment in Pakistan;
(ii) The port of entry in Pakistan;
(iii) The mode of transport from the exporting country to Pakistan;
(iv) The quantity of chemical(s) being imported; and
(v) Complete product safety information.
(vi) Transport of hazardous goods shall be subject to proper labeling as per the UN-adopted "Globally Harmonized System of Classification and Labeling of Chemicals" (GHS) Revision 6, 2015.

The reader should note carefully that while Schedule 17(2)(vi) requires labeling of hazardous goods in accordance with Revision 6 of the GHS, as will be discussed later, Pakistan does not mandate the use of GHS for SDS or other (non-transportation-related) labeling. The transport of hazardous substances is only permitted by a license.

In addition to the information contained in Form A of Schedule II, the applicant must also provide the following details:[637]

(i) The name and address of the person from whom the hazardous substance is to be collected;

(ii) The name and address of the person to whom the hazardous substance is to be delivered;

(iii) The quantity of hazardous substance to be transported;

(iv) The mode of transport, including full particulars and specifications of the motor vehicles or other conveyance;

(v) The route to be adopted between the origin and destination; and

(vi) The date and time of proposed transportation.

Schedule 1, Part I of the Hazardous Substances Rules contains a variety of definitions and chemical substance criteria to which many other sections of the Rules refer. For example, Schedule 1, Part I(a) provides a definition of only "flammable chemicals" and "toxic chemicals." The "flammable liquids" category includes "extremely flammable liquids," "very highly flammable liquids," and "flammable liquids."[638] Further, it segments the type of toxicity (extremely toxic, highly toxic, and toxic) based on corresponding oral toxicity (LD_{50} (mg/kg)), dermal toxicity (LD_{50} (mg/kg)), and inhalation toxicity (LC_{50} (mg/l)).[639]

Entities who possess test or similar flammability data for a substance, however, should be aware of some differences in definition. The Schedule lists "extremely flammable liquids" as chemicals which have flash points lower than or equal to 23°C and boiling points less than 35°C, "very highly flammable liquids" as chemicals which have flash points lower than or equal to 23°C and initial boiling points higher than 35°C, "highly flammable liquids" as chemicals which have flash points lower than or equal to 60°C but higher than 23°C, and "flammable liquids" as chemicals which have flash points higher than 60°C but lower than 90°C.[640] OSHA, in its "Flammable and Combustible Liquids" standard (29 C.F.R. § 1910.106; which has since changed the definition of "flammable" to align with GHS), previously only employed categories ranging from Class IA to Class IC for classification of flammable substances. All flammable liquids were Class I liquids, and specifically: Class IA liquids had flash points below 73°F (22.8°C) and boiling points below 100°F (37.8°C), Class IB liquids had flash points below 73°F (22.8°C) and boiling points above 100°F (37.8°C), and Class IC liquids had flash points at or above 73°F (22.8°C) and below 100°F (37.8°C).[641]

By the same token, entities that possess test or similar (e.g. read-across) toxicology data for a substance should be aware of some differences in definition. For example, as per the Schedule, a substance is "highly toxic" via an oral LD_{50} test in the appropriate animal test(s) at >5–50 mg/kg and "toxic" by the same measure at >50–200 mg/kg.[642] By comparison, OSHA considers a substance to be "highly toxic" via an oral LD_{50} test in the appropriate animal test(s) at <50 mg/kg and "toxic" by the same measure at 50–500 mg/kg.[643]

Schedule II of the Hazardous Substances Rules contains a 1,780 substance and compound list entitled "List of Prescribed Hazardous Substances." The list contains the name of the substance, the "name index" (often the same), the "CAS Sort Value" (the CAS number), and the "CAS/313 Category Codes." The term "CAS/313 Category

Codes" is, interestingly, not defined in the Hazardous Substances Rules. This may relate to Section 313 of the United States' Emergency Planning and Community Right-to-Know Act (EPCRA) legislation, under the Superfund Amendments and Re-authorization Act (SARA) of 1986. "SARA 313" is often a shorthand for the list of chemicals that facilities which fall under 29 C.F.R. § 1910.1200 must report on an annual basis if they meet or exceed applicable reporting thresholds.

Schedule III, "Isolated Storage at Installations Other Than Those Covered by Schedule 4," presents a list of substances and their associated threshold quantities (in tonnes) which "relate to each installation or group of installation[s] belonging to the same occupier where the distance between installations is not sufficient to avoid, in foreseeable circumstances, any aggravation of major accident hazards."[644] Schedule III, Part I ("Named Chemicals") also sets out a list of substances, their CAS number(s), and the threshold quantities; however, this list groups the substances by category (e.g. Group 1 – Toxic Substances, Group 3 – Highly Reactive Substances, and so forth).[645]

Schedule 6, the final Schedule of the Hazardous Substances Rules, presents a list of the "Information to be Furnished Regarding Notification of a Major Accident." Schedule 6 is itself a hybrid list-and-form, containing both the items which should be included in the notification (e.g. "Date, shift and hour of the accident") and specific boxes to be completed for various data points (e.g. "Type of major accident," with boxes for "Explosion," "Fire," and "Emission of dangerous substance"). Recall that this notification is required by Schedule 4. Schedule 7 ("Information to be Furnished for the Notification of Sites") lays out the data points necessary to meet the Schedule 6 requirements, as discussed earlier, while Schedule 8 ("Information to be Furnished in a Safety Report") defines the scope of information required in Section 8.

Schedule 9, "Safety Data Sheets," will be discussed later in the appropriate section. Schedule 10, "Format for Maintaining Records of Hazardous Chemicals Imported," contains a list of the data points necessary to be compliant with Schedule 18. Schedule 11, "Details to be Furnished in the On-Site Emergency Plan," delineates the requirements set out in Schedule 13. Finally, Schedule 12 presents the "Details to be Furnished in the Off-Site Emergency Plan," as per Schedule 14.

Pesticide Regulations

Prior to 1971, the import of pesticides was carried out by the Pakistani Federal Government, and as such were distributed among farmers through the provinces by the Authority. In 1971, the Agricultural Pesticides Ordinance (Ordinance) was promulgated, which changed the manner in which pesticides were managed in the country. The Department of Plant Protection (DPP), a sub-department under the Ministry of National Food Security & Research (MNFSR) (formerly the Ministry of Food & Agriculture), has been regulating the import and standardization of pesticides since then.[646]

Generally speaking, quality control at the *distribution and sale stage* for pesticides is done by the provincial governments.[647] These entities have nine pesticide labs and 645 inspectors.[648] As will be seen later, the verification of the pesticide quality is executed through the network of inspectors and pesticide labs registered by the DPP, under the Ordinance.[649]

Presently, pesticides are granted registration in Pakistan via one of three main categories (with the associated registration form type): "Form-1: Local Registration Under Trade Name," "Form-16: Generic Name," and/or "Form-17: Registered Abroad."[650] According to a June 2016 MNFSR presentation, there have been 680 brand name registrations, with 135 active ingredients, granted; 5,568 generic name registrations, with 97 active ingredients, granted; and 1,531 registered-abroad registrations, with 111 active ingredients, granted, for a total of 7,779 registrations.[651]

The Ordinance defines a "pesticide" as:

> any substance or mixture of substances used or represented as a means for preventing, destroying, repelling, mitigating or controlling, directly or indirectly, any insect, fungus, bacterial organism, nematodes, virus, weed, rodent, or other plant, or animal pest; but does not include a substance which is a "drug" in the meaning of the Drugs Act, 1940.[652]

Section 4 of the Ordinance, "Pesticides to be Registered," establishes the formal requirement for pesticide registration, noting that "No person shall import, manufacture, formulate, sell, offer for sale, hold in stock for sale or in any manner advertise any brand of pesticide which has not been registered in the manner hereinafter provided." Section 5 lays out the procedural requirements for applying to the MNFSR for the "registration of the brand under such name as he may indicate in the application."[653] Section 5(3) establishes the requirement for an agent or legal entity, directing that a Pakistani agent or representative sign the application for registration when the applicant is not domiciled in Pakistan.[654] To assist with subsequent examinations of sections, it should be noted that, under the Ordinance, "brand" means "the trade name applied by the importer, manufacturer, formulator or vendor to the goods imported, manufactured or sold by him."[655] This is important to keep in mind, as the manner in which the MNFSR uses the term may be confusing to those more familiar with other global pesticide registration schemes.

Provided the MNFSR is satisfied that the brand would not "deceive or mislead the purchaser with respect to the guarantee relating to the pesticide of its ingredients or the method of its preparation" (Section 5(4)(a)), that the "guarantee relating to the pesticide or its ingredients is not the same as that of another registered brand [by the same manufacturer] or is not so similar thereto as to be likely to deceive" (Section 5(4)(b)), is "effective for the purpose for which it is sold or represented to be effective" (Section 5(4)(c)), or is "not generally detrimental or injurious to vegetation, except weeds, or to human or animal health, even when applied according to directions" (Section 5(4)(d)), it is likely to consider granting registration.[656] As per Section 6 of the Ordinance, "the registration of a brand of a pesticide shall be effective from the date of its registration until the thirtieth day of June of the third year following the year of registration."[657] The subsequent section of the Ordinance lays out the conditions under which the MNFSR may, after affording the registrant an opportunity for a hearing, cancel a granted registration.

Section 12 sets out the group whose function is to "advise the [Federal Government] on technical matters arising out of the administration of this Ordinance," as well as any other assigned functions.[658] The group, the Agricultural Pesticide Technical Advisory Committee, will consist of a Chairman, Vice-Chairmen [plural], and other members (specified as officers of the Federal Government or a Provincial Government or persons "representing trade and industry engaged in pesticide business"), such as the Federal Government may appoint.[659] Sections 12(4)–12(10) further detail the terms, processes, and powers of the Committee.

In Section 13, the "Pesticide Laboratory" is established for the "submission of samples for analysis or test." Interestingly, the management of CBI relating to pesticide testing is addressed in this section, albeit in somewhat vague terms, requiring that the "formulae of brands of pesticides samples… shall be duly safeguarded."[660] Further, "local performance of environmental safety tests is not required to be conducted by the person registering a pesticide, and results of tests conducted outside Pakistan may be submitted."[661] Careful readers may recall that this is counter to how some other Middle Eastern countries, such as Israel, manage their efficacy and environmental safety testing requirements.

Section 16 authorizes an inspector to, subject to his jurisdiction, "enter upon any premises where pesticides are kept or stored… including premises belonging to a bailee, such as a railway, a shipping company or any other carrier, and may take samples therefrom for examination," while Section 17 sets forth the specific requirements for obtaining, documenting, and managing the sample(s).

Remarkably, Section 20 of the Ordinance specifically provides that "any person who has purchased a pesticide" has the statutory ability to apply to the government to have the product tested and/or otherwise analyzed, and to receive a copy of the result(s).[662] This provision would appear to reasonably extend to consumers and potential competitors. Sections 21–28 detail the penalties which may be applied under the Ordinance.

Section 29 contains the provisions under which rules may be (further) developed with respect to the nomenclature for pests, how an application shall be made and the associated fees, the "language of the tags or label to be attached to the containers and packages," as well as "the character and location of the printing to be marked on such tags, labels and containers," "the pesticides that are generally detrimental or injurious… even when used according to directions," the "pesticides that are to be labelled 'Poison' and their antidotes," as well as other items.[663]

Pakistan maintains a list of 26 pesticide active ingredients which are banned from use in the country. These are:[664]

1. BHC
2. Binapacryl
3. Bromophos Ethyl
4. Captafol
5. Chlordimeform
6. Chlorobenzilate
7. Chlorthiophos
8. Cyhexatin
9. Dalapon
10. DDT
11. Dibromochloropropane and Dibromochloropropene
12. Dicrotophos
13. Dieldrin
14. Disulfoton
15. Endosulfan

16. Endrin
17. Ethylenedichloride and Carbontetrachloride
18. Heptachlor
19. Leptophos
20. Mercury Compound
21. Methamidophos
22. Methyl Parathion
23. Mevinphos
24. Monocrotophos
25. Toxaphene
26. Zineb

Returning to the three types of registration discussed earlier ("Form-1: Local Registration Under Trade Name," "Form-16: Generic Name," and "Form-17: Registered Abroad"), the following is a summary of the registration process under each type:[665]

1. REGISTRATION PROCESS UNDER FORM-1: (TRADE NAME)
   1. Applications are submitted on prescribed Form-1 along with registration fee & samples;
   2. Product sample is analyzed by the FPTRL [Federal Pesticide Testing and Reference Laboratory] and, after satisfactory report, sent to the Provinces for conducting field trials;
   3. Efficacy trials at different Federal/Provincial Research Institutes for two crop seasons by at least two Research Institutes;
   4. Standardization of the product by the Provincial Standardization Committee and recommendations forwarded to the Department of Plant Protection;
   5. Examination by the Agricultural Pesticides Technical Advisory Sub-Committee, and then approval by the Agricultural Pesticides Technical Advisory Committee; and
   6. Granting a certificate of registration for a period of three years or shorter.
   The following documents are required from the manufacturer:[666]
   1. A letter of authorization:
      a. From the manufacturer (in original) with name, designation, seal & signature, telephone, fax and email address; and
      b. For local manufacturing, a letter of authorization (in original) for the technical grade material should be provided from the manufacturer with all the information.
   2. 100% break up [full composition]/recipe and specification of formulated product:
      a. For imported product, the recipe and specification should be analyzed by an internationally accredited lab/GLP [Good Laboratory Practices] lab; and
      b. For local manufacturing, 100% break up/recipe and specification analyzed by internationally accredited lab/DPP or labs notified by DPP.
   3. 100% break up/composition and specification of technical grade material:
      a. Toxicological, eco-toxicological, stability, MRL data (TC) from internationally accredited lab/GLP lab; and
      b. For formulated product toxicological study data & PHI [Pre Harvest Interval] period only from an internationally accredited lab/GLP lab.
   4. A registration certificate of the formulated product or proof of manufacturing;

5. A registration certificate of technical grade material of the pesticides or proof of Manufacturing [sic]; and

6. A sample of formulated product (1 kg/l), TC (0.5–1 kg) and analytical standard (2–5 g). (A sealed sample should be submitted to the FPTRL directly from the manufacturer.)

2. IMPORT PERMISSION OF PESTICIDES [sic] UNDER THE GENERIC SCHEME (FORM-16)[667]

1. An application should be made on the prescribed Form-16, along with the necessary fee & samples;

2. Sample analysis by FPTRL-DPP; and

3. An import permission certificate for a period of three years or shorter will be granted upon approval.

The following documents are required from the manufacturer:[668]

1. An authorization letter from the manufacturer in the country of origin;

2. A 100% break up/recipe and specification of formulated product by an internationally accredited lab/GLP lab;

3. A 100% break up/composition and specification of technical grade material as registered with registration authority in the country of origin;

4. MRL data & PHI period of the product by an internationally accredited lab/GLP lab;

5. Toxicological study data of the product by an internationally accredited lab/GLP lab;

6. A registration certificate of the formulated product/technical grade material, or proof of manufacturing issued by the registration authority in the country of origin; and

7. A sample of formulated product (1 kg/l), TC (0.5–1 kg) and an analytical standard (2–5 g) should be submitted to the FPTRL directly from the manufacturer.

3. IMPORT PERMISSION OF PESTICIDES [sic] REGISTERED IN THE COUNTRY OF MANUFACTURE (FORM-17)[669]

1. An application on the prescribed Form (Form-17), along with the registration fee for import of pesticides registered in the country of manufacture, and not registered under Form-1 or Form-16 with [the] following conditions:

   a. Documented proof of the pesticide's registration in the country of manufacture;

   b. Proof of use of the pesticide in any member country of the Organization for Economic Cooperation and Development (OECD), China or India; and

   c. Documentary proof of the pesticide's extensive use on relevant crop(s) and its pest(s) in the country of origin or any other country specified in clause (b).

The following documents are required from the manufacturer:[670]

1. Documentary proof of registration of the formulated product under the brand name and technical grade material of the product in the country of origin;

2. Documentary proof of the use of the pesticides in OECD countries, China or India;

3. Documentary proof of extensive use against pests of the crop in the country of origin, OECD countries, China or India;

4. An authorization letter from the manufacturer in the country of origin;

5. A 100% break up/recipe and specification of the formulated product analyzed by an internationally accredited lab/GLP lab;

6. 100% break up/composition and specification of technical grade material in accordance as registered with [the] registration authority in the country of origin;
7. MRL data and PHI period of the product from internationally accredited lab/GLP lab; and
8. Toxicological study data of the formulated product and TC from internationally accredited lab/GLP lab.

In order to facilitate the registration and/or import of pesticide products, the DPP introduced the "Fast Track Registration of Agriculture [sic] Pesticide" program through S.R.O. 1017(I)/2014, effective November 5, 2014. This system applies to the products already registered via Form-16 and Form-17.

Furthermore, to verify and ensure the quality of imported pesticides, as discussed briefly earlier, the DPP has established five "Pre-Shipment International Inspection Agencies."[671] These agencies verify product quality by obtaining an original copy of the analysis report from the internationally accredited/GLP laboratory employed, as well as undertaking a random sampling of 10% of the pesticides arriving at port(s). Samples of pesticide(s) are taken and analyzed in the Provincial Pesticide Laboratories. The sample is split, with one portion given to the vendor and the other sent to the FPTRL, DPP-Karachi.[672] Further, the inspectors visit stores and shops to inspect pesticides, to ensure the sale of high-quality, effective, and approved products. In a case where a sample is declared unfit by the provincial laboratory, a First Information Report (FIR) is registered, and a court case is processed for trial.[673] According to the website of the Punjab Police, an FIR is defined as "an account of a cognizable (i.e. over which police has jurisdiction) offence that is entered in a particular format in a register at the police station."[674] The firm and/or vendor has the right to submit an appeal to the DPP within one month, for retest of the other sealed portion of the sample.[675]

Occupational Safety and Health Regulations

"There is no independent legislation on occupational safety and health issues in Pakistan."[676] The Hazardous Substances Rules discussed earlier with respect to the province of Sindh do contain components of such, although they are relatively limited in scope and, as noted, only applicable to the province of Sindh.

As such, OSH is managed under several different pieces of Pakistani legislation, such as the Factories Act (1934), specifically under Chapter 3.[677] Chapter 3 of the Act has general provisions on safety and health in the workplace. Provincial governments are permitted to make rules under this Act, and inspectors also have discretion in defining the rules. As such, all the provinces have devised what have been termed "Factories Rules." Similarly, Chapter 5 of the Mines Act provides for various safety and health arrangements.[678]

Chapter 3 discusses various safety aspects and requirements for areas including, but not limited to:[679]

- Cleanliness;
- Disposal of wastes and effluents;
- Ventilation and temperature;
- Lighting

- Drinking water;
- Precautions against contagious or infectious disease;
- Work on or near machinery in motion;
- Floors, stairs, and means of access;
- Protection of eyes; and
- Explosive or inflammable dust and/or gas.

The Hazardous Occupations Rules, promulgated in 1963 under the authority of the Factories Act, also refer tangentially to OSH management in the country. These rules not only specify some hazardous occupations, but also authorize the Chief Inspector of Factories to declare any other process as hazardous.[680] Other laws with an OSH aspect include: the Dock Laborers Act (1934), the Mines Act (1923), the Workmen Compensation Act (1923), the Provincial Employees Social Security Ordinance (1965), the West Pakistan Shops and Establishments Ordinance (1969), and the Boilers and Pressure Vessels Ordinance (2002).[681] All of the foregoing regulations require the appropriate government (Federal or Provincial) to appoint qualified individuals as inspectors, whose duty is to enforce these laws. The usual powers of inspectors include the right to enter and inspect any workplace, and to take items as evidence where applicable.[682]

Of particular note is that various Ministries and organizations maintain data with respect to OSH (e.g. accidents); however, no centralized database or similar repository exists for broader study and strategic planning of such incidents.[683] This is perhaps emblematic of a larger issue within the Pakistani chemical industry: "Many industries neither maintain records of accidents, injuries, and fatalities, nor do they report to public authorities."[684]

The Center for Improvement of Working Conditions and Environment (CIWCE) is a pioneering institution in Pakistan (working under the Directorate of Labor Welfare, Punjab), which provides training, information, and research facilities for the promotion of safety, health, and a better working environment in industries and businesses. The goal is to improve the reporting, tabulating, and management processes.[685]

Waste Regulations

The Hazardous Substances Rules (Rules), discussed earlier, refer in Section 5(2)(b) to the requirement by entities licensed under the Rules to establish and maintain a Waste Management Plan (Plan). The specific requirements for this Plan are laid out in Section 19 of the Rules. Readers should recall, however, that the Rules only refer to the management of those substances which are defined as "hazardous." That is not to say that other waste types – such as municipal, non-hazardous, industrial, and so forth – are not regulated, but rather the Rules relate only to this specific category.

The Plan shall encompass two key aspects. First, it shall "provide for the generation, collection, transport and disposal of the hazardous waste in a manner which shall protect against an adverse environmental effect" and second, it shall "ensure that the hazardous waste is not mixed with non-hazardous waste, unless the applicant can prove that such mixing will better protect against an adverse environmental effect."[686] The Plan must be reviewed every year by the licensee, taking into the evaluation if there are any new technologies and management practices which can better protect against the adverse environmental effect than those identified originally/previously, and if a revised waste management plan and new EIA should be submitted with the license renewal application.[687] The licensee is obligated under Section 19 to inform PAK-EPA at least yearly of the quantity of characteristics of the hazardous waste which they generated in the previous year, and also the progress regarding the implementation of the waste management plan.[688]

Safety Data Sheets and Labels

There exists no overarching regulation or provision governing the use of SDSs in Pakistan. While not legally binding, however, International Chemical Safety Cards may be used.[689] As an aside, English and Urdu versions of these Cards are maintained by the CIWCE. EU SDS – in either Urdu or English – are accepted.[690] In the past, Pakistan has asked for technical assistance regarding the GHS, but no real effort has been undertaken to develop or implement such a system in the country.

Readers should note that Schedule 9 of the Hazardous Substances Rules, "Safety Data Sheet," does present a form which may be completed, with the section headings and data fields appearing very similar to both the ANSI Z400.1-1993 and the United States' 2012 "Hazard Communication Standard" regulation.

Similarly, with respect to chemical substance labeling, Pakistan does not have a universal system for labeling or marking requirements on products, however there are a few industry-specific regulations.[691] "Other than… some labeling requirements for chemical pesticides and pharmaceuticals, there are no provisions for labeling industrial chemicals. However, for the sake of worker safety, many companies in Pakistan have decided to label chemical containers pursuant to other chemical standards." Any labels employed should contain all specific hazard text provided in Urdu or English.[692] As above, Schedule 9 of the Hazardous Substances Rules, Sections 1 and 2, does refer to this topic, but again only in generalities.

8

Saudi Arabia

National Overview

The modern state of the Kingdom of Saudi Arabia (Saudi Arabia), bordered by Iraq, Jordan, Kuwait, Oman, Qatar, the United Arab Emirates (UAE), and Yemen, and which is approximately a fifth the size of the United States, was founded on December 23, 1932 by ABD AL-AZIZ bin Abd al-Rahman Al SAUD (Ibn Saud).[693] Saudi Arabia is the most influential member of the GCC.[694]

The country is a leading producer of oil and natural gas, and holds about 16% of the world's proven oil reserves as of 2015.[695] The country is comprised of 13 provinces (*mintaqah*): Al Bahah, Al Hudud ash Shamaliyah (Northern Border), Al Jawf, Al Madinah (Medina), Al Qasim, Ar Riyad (Riyadh), Ash Sharqiyah (Eastern), Asir, Ha'il, Jazan, Makkah (Mecca), Najran, and Tabuk.[696]

> Saudi business [sic] and laws still favor Saudi citizens, and Saudi Arabia still has trade barriers, mainly regulatory and bureaucratic practices, which restrict the level of trade and investment. For example, only Saudi nationals are permitted to engage in trading activities, and only Saudis are permitted to register as commercial agents.[697]

Further, while entities exporting to Saudi Arabia are not required to establish a local Saudi agent or distributor to sell products to Saudi companies, commercial regulations in the country restrict imports for resale and direct commercial marketing to Saudi nationals and wholly Saudi-owned companies.[698]

Importers should bear in mind that no Israeli commodities or products of any kind shall be brought in or imported into Saudi Arabia.[699] Further, goods and commodities made in Israel, having any percentage of their composition from Israel, or reshipped from Israel will be considered Israeli, whether such goods and commodities come directly or indirectly from Israel.[700]

Saudi Arabia has traditionally followed the Islamic (or Hejira) calendar, but in December 2016 they announced a transition to the Gregorian calendar. Until the transition is complete, observers will likely see both calendars in use.[701]

Regulations are published in the Official Gazette of Saudi Arabia, *Umm al-Qurā* (www.uqn.gov.sa/).

Chemical Regulation in the Middle East, First Edition. Michael S. Wenk.
© 2018 John Wiley & Sons Ltd. Published 2018 by John Wiley & Sons Ltd.

Governmental Structure

Under the Saudi Arabian governmental system (an absolute monarchy), the King is also the Prime Minister: both chief of state and head of the executive branch. He appoints his cabinet, the Council of Ministers, every four years. The King rules by Royal Decrees, which are issued in conjunction with the Council of Ministers. The Council of Ministers formulates and supervises the implementation of governmental policy, as well as overseeing the resolutions passed by the Consultative Council, *Majlis Al-Shura*, whose purpose is to advise the King and the Council of Ministers on matters related to government programs and policies.[702] There are no elections, as the monarchy is hereditary.[703]

Each province is headed by an Emir appointed by the King. The Emir is assisted by a provincial council, encompassing the heads of the province's governmental departments, and a ten-member council of prominent individuals in the community who are appointed to four-year, renewable terms.[704] The Emir answers to the Ministry of the Interior.[705]

The legislative branch – the Consultative Council – is comprised of 150 seats, with its members appointed by the monarch to four-year terms.[706] The legal system is Islamic (*sharia*) law, with some elements of Egyptian, French, and customary law.[707] The judicial branch is comprised of the High Court (which consists of the Court Chief, and is organized into circuits with three-judge panels), the Court of Appeals, a Specialized Criminal Court, first-degree courts composed of general, criminal, personal status, and commercial courts, a Labor Court, and a hierarchy of administrative courts.[708]

Commercial and business disputes are managed by special administrative bodies. They include the Board of Grievances, and a number of commissions and committees, which hear cases involving particular subject matters such as labor or commercial disputes.[709] "In addition, the Commission for the Settlement of Commercial Disputes hears cases involving disputes between companies. Decisions of the Commission may be appealed to a special appeals tribunal."[710]

Key Chemical Regulatory Agencies

Within Saudi Arabia, the Saudi Arabian Standards Organization (SASO) is responsible for the promulgation of a wide range of standards for the country.[711] Relative to chemical regulations, SASO puts forth national standards for products, testing methods, safety measures, and environmental testing, among others.[712] In fact, SASO "is the only Saudi organization responsible for setting national standards for commodities and products, measurements, testing methods, metrological symbols and terminology, commodity definitions, safety measures, and environmental testing, as well as other subjects approved by the organization's Board of Directors."[713] Although SASO holds an advisory, as opposed to an executive, role, "it coordinates its activities among different executing agencies in the country to control product quality and standards."[714] As of mid-2016, SASO had approximately 20,500 standards.[715]

The Saudi Ministry of Commerce and Industry has responsibility for the issuance of import licenses for commercial chemicals (except for explosives and chemicals considered "dangerous," which must be approved by the Ministry of Interior and/or the Ministry of Health).[716] The Ministry of Industry and Electricity is responsible for issuing licenses for chemicals which are imported by the national industries (here, explosives and chemicals considered "dangerous" must be approved by the Ministry of Commerce and Industry).[717]

The President of Meteorology and Environment (PME), previously known as the Meteorology and Environmental Protection Administration (MEPA), is the primary agency in Saudi Arabia with responsibility for environmental issues. Among the responsibilities of the PME are:[718]

1. Environmental surveys and pollution assessment and control;
2. Establishment of environmental standards and regulations;
3. Recommendations on practical measures for emergency situations;
4. Keeping up to date with environmental developments internationally; and
5. Preparing and issuing climatological, environmental, and meteorological analyses, forecasts, and bulletins.

The Third Article of the "Law of Chemicals Import and Management," also known as Decree No. M/38, which entered into force on June 12, 2006 (Law), lays out the Ministries which have responsibility for chemical management:[719]

a. Ministry of Interior: Chemicals used in explosives;
b. Ministry of Higher Education: Chemicals used in educational institutions;
c. Ministry of Health: Chemicals used in the preparation of medicines as well as chemicals needed in the health sector, including non-radioactive reagents;
d. Ministry of Agriculture: Chemicals used in the preparation of pesticides, soil enhancers, fertilizers, veterinary medicines, and chemicals needed in agricultural research centers;
e. Ministry of Water and Electricity: Chemicals used in water and sewage treatment and their plants, chemicals needed in water and sewage research laboratories and centers, as well as electricity companies; and
f. Ministry of Commerce and Industry: Chemicals traded in local markets and imported by commercial establishments and companies, as well as chemicals used by petroleum and mining factories and companies.

The Third Article goes on to note that "Ministries referred to in paragraphs (c, d, e and f) above many not issue permits of import or clearance except by agreement with the Ministry of Interior."[720] Article 12 directs the above competent agencies, "each within their jurisdiction [to] undertake the following":[721]

1. Monitor and inspect entities and facilities handling chemicals to ensure their compliance with provisions of this Law and its Regulations and all directives issued relating to chemicals; and
2. Record and establish violations of provisions of this Law and prepare minutes thereon. The Regulations shall stipulate procedures for recording and establishing violations.

The Law assigns the following roles and responsibilities to the Ministry of Interior in Article 8:[722]

• List and monitor all imported chemicals used in explosives, and importers thereof;
• Escort trucks carrying hazardous chemicals – specified by the Regulations – to ensure their safety on [the] road;

- Set up safety and protection measures against chemical hazards and oversee their implementation; and
- Form intervention teams for chemical accidents, and provide such teams with [the] training and equipment necessary to perform their tasks.

In a related vein, Article 9 assigns the following roles and responsibilities to the customs authority:[723]

- Complete customs procedures necessary for any shipment of chemicals coming into the Kingdom upon submission of the clearance permit;
- Coordinate with the competent agencies to overcome any hindrance or delay in clearance of imported chemicals; and
- Notify the Presidency of Meteorology and Environment of unclaimed chemicals.

Finally, the responsibilities of the Presidency of Meteorology and Environment are delineated in Article 10:[724]

1. Coordinate with concerned agencies to establish database [sic] containing:
   a. A list of chemicals cleared, importing agencies, and chemicals destroyed;
   b. A list of prohibited chemicals strictly not permitted to enter the Kingdom and restricted chemicals permitted only in accordance with special conditions and instructions, and a copy thereof shall be provided to the agencies concerned; and
   c. All that relates to chemical waste, its characteristics, hazard level, and appropriate circumstances for its storage, transport and recycling, as well as methods of storage and disposal thereof.
2. Regulate means of sensing and warning of chemical accident hazards, and devise plans to face emergency cases affecting the environment, in coordination with relevant agencies;
3. Issue permits for building plants and facilities for chemical waste treatment in accordance with standards specified by the Regulations, and monitor such plants and facilities as well as their disposal;
4. Establish rules and procedures to control processes of destroying and disposing of chemical waste, and monitor the implementation of such rules and procedures, in line with relate[d] laws and treaties;
5. Coordinate with [the] relevant agencies to provide appropriate sites for destroying and dumping of chemicals, and oversee the establishment of burial sites and the destruction and dumping process; and
6. Take necessary measures regarding chemicals with agencies seeking to dispose of them, upon notification of the Presidency of Meteorology and Environment thereof [sic].

Key Chemical Substance Regulations

"The Constitution of Saudi Arabia, under Article 32, sets forth the requirement for the country to preserve, protect, and improve the environment."[725] No specific criteria for achieving this requirement are provided in the Basic Law, which is the implementing regulation of the Constitution.[726]

In Saudi Arabia, there is currently a system in place to regulate import of chemicals and there are efforts to coordinate different chemicals conventions. Further, Saudi Arabia has a national program on chemical safety involving all related authorities. The rules and regulations are stipulated by general environmental regulations dealing with the waste of chemical substances. There are a number of industry measures to ensure chemical safety, in cooperation with the Ministry of Commerce and Trade. This Ministry deals with regulating chemical imports.[727]

As has been noted briefly, and will be examined in more detail later, chemicals imported by commercial importers require an import license granted by the Ministry of Commerce, while chemicals imported by national industries require both an approval and an import license from the Ministry of Industry and Electricity. In addition, the Ministry of Agriculture and Water issues import licenses for, among other items, seeds and fertilizers, pesticides, feed additives, and biological materials.[728]

Saudi Arabia does not publish an inventory of chemicals and does not have a new chemical notification scheme.[729] In fact, "Any chemicals not subject to the ministries [sic] of Interior and Health shall not be refused release as long as the information and documents are complete."[730] That said, however, "the more sophisticated legislative and regulatory requirements pertaining to the chemical industry exist primarily on a local level [in the country]."[731] For example, the Saudi cities of Jubail and Yanbu presently have their own regulatory systems in place to manage chemicals.[732] This is a still relatively uncommon situation in the Middle East, although several cities in the UAE (e.g. Abu Dhabi and Dubai), as well as the Pakistani province of Sindh, have undertaken similar actions. "Since inception… [it] has been determined that Jubail and Yanbu would be models of environmental planning and management in addition to being productive manufacturing centers."[733] Indeed:

> The Royal Commission for Jubail and Yanbu has issued the "Royal Commission Environmental Regulations (RCER)" to be adopted by industries both in Jubail and Yanbu. Any facility operating or planning to operate on the Royal Commission property will be required to comply with these regulations. These regulations will be updated periodically to reflect the environmental needs of the cities and the latest in pollution control technologies.[734]

Section 1.5.4 of the RCER expressly recognizes that, when a given environmental regulation ("practice") does not exist at the Saudi Arabian national level, "the Royal Commission will reference other recognized regulations as a basis for technical justification or establishment of a change in the following order:"[735]

a. Saudi National/PME Standards;
b. U.S. Environmental Protection Agency (US EPA);
c. U.S. State environmental protection rules and guidelines;
d. European Union Members' environmental rules and guidelines; and
e. Other internationally recognized and accepted regulatory bodies.[736]

Looking at three examples of such "establishments of change," we see that the RCER has established its own Safety Data Sheet (SDS) and its own product labeling requirements, as well as a chemical inventory. With respect to SDSs, the requirements include an obligation for facility operators in the cities to retain copies of current SDSs, for all hazardous materials present at the facility.[737] While there is no specification for the format (sections) of the SDS, the document(s) must be presented in <u>both</u> Arabic and English.[738] For chemical product labeling, there are three key aspects to note:[739]

- Containers holding hazardous materials must be individually labeled to reflect the actual contents of the container;
    - The label must include:
        o Contents and associated hazards (according to the UN Chemicals Hazard Classification System), or
        o A unique identification that is cross-referenced to a document which lists the contents and the hazards.
- Re-labeling of materials whose original labels have been obliterated or lost must be conducted with care to avoid mislabeling; and
- Unidentified substances must be tested or analyzed to confirm the identity of the material, prior to re-labeling.

Finally, the RCER has put forth a "quasi" chemical inventory, in the sense that particular chemicals must be tracked (inventoried) at facilities, but such is not an "inventory" in the same way that, for example, the U.S. TSCA Inventory is. Generally speaking, under TSCA, companies wishing to manufacture or import chemical substances into the United States must ensure that their substance is listed on the TSCA Inventory. Under the RCER requirement, facility operators must keep an inventory of hazardous chemicals that exceed 50 kg for highly toxic chemicals, and 5,000 kg for other chemicals.[740] Such an inventory must include the following information:[741]

1. The chemical name and trade name;
2. The chemical composition, including concentration of hazardous components;
3. The physical form of the material;
4. Temperature (°C) and true vapor pressure (kPa) for liquids and gases;
5. Storage quantity (annual average and maximum values);
6. Associated hazard classification; and
7. End use.

Further, an annual report of the inventory must be filed with the Royal Commission.

Saudi Arabia promulgated the Environmental Law, enacted on September 24, 2001, in the form of the Council of Ministers Resolution No. 193.[742] It entered into force on October 31, 2002, and its Implementing Rules were published on September 30, 2003.[743] The Environmental Law is the overarching law for environmental regulation in the Kingdom. The Law's purposes are specified in Article 2 as:[744]

1. Conserve, protect and develop the environment and prevent pollution thereof;
2. Protect public health from hazards resulting from activities and acts harmful to the environment;
3. Conserve and develop natural resources and promote awareness of proper use thereof;
4. Make environmental planning an integral part of overall planning for development of industrial, agricultural, urban and other fields; and

5. Raise awareness of environmental issues, promote the sense of individual and collective responsibility to conserve and improve the environment and encourage national voluntary efforts in this field.

Under its auspices, the MEPA is designated as the competent authority, and the (perhaps strangely) Minister of Defense and Aviation and Inspector General is designated as the competent Minister.[745] Article 1 defines "environment" extremely broadly, and perhaps uniquely when compared with various other global regulatory definitions of the same as "All that surrounds man whether water, air, land and outer space as well as contents thereof: nonliving objects, plants, animals, different forms of energy, systems, natural processes and human activity."[746]

Of particular interest within the Environmental Law are the following provisions and duties of the competent authority (with their corresponding Articles): "Review and assess the condition of the environment, develop monitoring means and tools… document and publish information on the environment… set, issue, review, develop and construe environmental protection standards… draft environmental laws relating to its responsibilities… promote environmental awareness on all levels."[747] Article 4 specifically directs "licensing authorities" to "ensure that environmental assessment studies are conducted in the feasibility study phase for projects with potential adverse impact on the environment," while the responsibility to conduct the studies themselves falls to the entity proposing the project.[748] Of specific note is Article 6, which requires the use of Best Available Technologies (BAT) and the "least polluting materials" for specified project categories.[749] There is no language in the Article referring to any type of cost–benefit balance, however.

Another unique aspect of the Environmental Law, at least when compared with other countries' environmental regulations, is the requirements in Article 7 to include "environmental concepts into curriculums of various stages of education," as well as to "promote environmental awareness programs in the media and reinforce the concept of environment protection from an Islamic perspective."[750] While the latter is not specifically defined in the Article, the third subsection does direct that such "shall promote the role of the mosque in encouraging society to conserve and protect the environment."[751]

Article 8 contains a statutory mandate to "conserve and develop renewable resources and prolong the lifespan of non-renewable resources," as well as to "achieve harmony" between use rates and capacity of natural resources and, again somewhat uniquely, to "develop technologies of conventional building materials."[752] Article 12 of the Environmental Law touches briefly on an aspect which is normally enshrined in worker and workplace statutes, declaring "An enclosed and semi-enclosed public place shall meet ventilation requirements adequate to its area, capacity and type of activity carried out therein."[753]

As briefly discussed earlier, with respect to chemicals, the "Law of Chemicals Import and Management" (Decree No. M/38, June 12, 2006) (Law) refers to "all that has to do with chemicals including their production, manufacture, circulation, transportation, storage, treatment, damage, and dumping."[754] The Law defines chemicals very broadly, and with a surprisingly circular definition, in the first Article as "[a]ny chemical substance whether gaseous, liquid or solid."[755] Many of the definitions listed in Article 1 are extremely general or circular, such as "Accumulated Chemicals: Chemicals stored for an extensive period of time…" and "Chemical Waste: Chemical waste to be disposed of…"[756] Exempt from the purview of the Law via Article 16 are "Medicines, Chemicals for household use as specified by the Regulations, [and] Chemicals imported directly by the armed forces for military purposes."[757]

The crux of Saudi Arabia's chemical import regulation derives from Article 2 of the Law, which states that "Chemicals may not be imported without obtaining a permit, nor may they be cleared by customs without a clearance permit."[758] Article 4 of the Law delineates that "Agencies [entities subject to the permit requirements of the Law] concerned shall provide the Presidency of Meteorology and Environment with a copy of the chemicals clearance permit upon issuance."[759]

The entities charged with granting an Article 2 permit are numerous and, as with many of the countries examined previously, each has its own purview. As such, applicants may discover the need to hold discussions with, and potentially apply for multiple permits from, more than one Ministry. The specific Ministries involved in permit issuance, as per Article 3, are:[760]

a. Ministry of Interior: Chemicals used in explosives;
b. Ministry of Higher Education: Chemicals used in educational institutions;
c. Ministry of Health: Chemicals used in [the] preparation of medicines as well as chemicals needed in the health sector, including non-radioactive reagents;
d. Ministry of Agriculture: Chemicals used in [the] preparation of pesticides, soil enhancers, fertilizers, veterinary medicines and chemicals needed in agricultural research centers;
e. Ministry of Water and Electricity: Chemicals used in water and sewage treatment and their plants, chemicals needed in water and sewage research laboratories and centers, as well as electricity companies; and
f. Ministry of Commerce and Industry: Chemicals traded in local markets and imported by commercial establishments and companies, as well as chemicals used by petroleum and mining factories and companies.

Potential applicants should note that the Ministries of Health, Agriculture, Water and Electricity, and Commerce and Industry, may not issue permits for either import or clearance except by agreement with the Ministry of Interior.[761] Further, once a permit is granted by a Ministry, Article 4 requires that such Ministry shall provide the Presidency of Meteorology and Environment with a copy of the permit.[762] Article 5, discussed later, refers to the requirements relating to chemical containers and labels. Articles 6 through 8 address the various administrative and procedural requirements for permit issuance.

Interestingly, Article 7 requires three items which are not normally found as governmental authority responsibilities in various global chemical management statutes, but which fall therein to the Ministry of Interior: to escort trucks transporting hazardous chemicals (as defined in the Law) to ensure safety, to devise "safety and protection measures against chemical hazards and oversee their implementation," and to create "intervention teams" for chemical accidents, as well as providing the same with training and equipment.[763] While one may argue that the third responsibility is similar to the emergency response or hazardous materials ("HAZMAT") teams present in many major U.S. cities (among other locations), the other two items are often seen as responsibilities of individual companies. Article 8 directs the customs authority to "coordinate with the competent agencies to overcome any hindrance or delay in clearance of imported chemicals."[764] Generally speaking, such "overcoming" of delays related to the importation of chemicals falls to the private entity and/or his/her customs broker or local agent. In this Article, the Saudi government is effectively taking this role, perhaps illustrating the importance that the government places on the import of chemicals into the Kingdom.

Article 9 of the Law enumerates the responsibilities of the Presidency of Meteorology and the Environment in substantial detail. The entity shall coordinate with the various concerned Ministries identified previously to establish a database which contains:[765]

a. A list of chemicals which have been cleared for entry into Saudi Arabia, as well as the importing agencies (Ministries). In addition, the database should contain a list of all chemicals destroyed;
b. A list of both prohibited and restricted chemicals, with a copy of [the] same to be distributed to the agencies concerned;
c. "All that relates to" chemical waste: the characteristics and hazard level thereof, proper means for storage, transport and recycling of the waste, as well as for its disposal;
d. A means of identifying when chemical accidents occur, as well as plans to respond to such emergency situations;
e. The means to issue permits to construct facilities for chemical waste treatments, as well as to monitor [the] same;
f. Rules and procedures to manage the disposal and destruction of chemical waste;
g. Processes for inter-Ministry coordination to establish and maintain waste disposal sites, as well as to oversee the destruction and dumping process; and
h. The necessary measures to be taken regarding chemicals for agencies seeking to dispose of them, upon notification from the Presidency of Meteorology and Environment.

Article 10 of the Law sets out the requirements with which "chemicals importers and management" [sic] shall comply, to lawfully import and use such substances in Saudi Arabia. These are:[766]

a. The completion of customs clearance procedures prior to or upon arrival of the chemicals into Saudi Arabia;
b. The collection and transport of all imported chemicals within three working days from the date the substances clear customs;
c. Safely transport chemicals by licensed means of transportation;
d. Notify the Ministry of Interior prior to transporting hazardous chemicals;
e. Store chemicals in designated sites, in accordance with conditions and instructions as stipulated;
f. Ensure that workers managing such substances are properly trained in their safe handling;
g. Apply standard specifications to chemical substance containers;
h. Refrain from use of imported chemicals for purposes other than those for which they are intended and/or permitted, except with the approval of the competent agency, or their use outside of designated site(s);
i. Ensure chemicals are handled only by individuals qualified in the field of safety and accident prevention;

j. Inform the agency concerned, and the Presidency of Meteorology and Environment, periodically, of accumulated or expired chemicals or chemical waste in their possession; and

k. Dispose of chemical waste only by a specialized licensed facility, and furnish the Presidency of Meteorology and Environment with proof of such proper disposal.

Article 13 of the Law lays out the penalties, "without prejudice to any severer punishment provided for in other laws," for non-compliance with the above.[767] These are set out as "A fine not exceeding five hundred thousand riyals (SAR). Imprisonment for a period not exceeding five years. Debarment from chemicals import and management for a period not exceeding five years."[768] While financially significant, the same Article also provides that "in addition to the aforementioned punishments, a judgement may be issued to return imported chemicals in question to their point of origin or destroy them at the violator's expense."[769] In all cases, the violator shall be compelled to remove and remedy the cause of the violation.[770] The Board of Grievances is granted the authority as per Article 15 to have jurisdiction, and to "decide all violations, disputes, and claims for compensation arising from the implementation of the provisions of this Law."[771]

On March 24, 2012, the "Technical Guideline of 2012 on the Prevention of Major Accidents" (Technical Guideline) entered into national law. The Technical Guideline directs those entities involved in activities which relate to the "manufacturing, processing, using, storing, or otherwise handling of dangerous substances," regardless of the size of the operation(s) or the location(s), to take all actions necessary to prevent major accidents.[772] However, threshold values do apply. Examples of such "major accidents" are the release of toxic materials, the release of flammable materials, fires, explosions, major structural failures, and any accident that involves dangerous substances.[773]

Pesticide Regulations

As of the close of 2015, there were in excess of 50 companies licensed to import agricultural pesticides into Saudi Arabia.[774] In addition, there were more than 300 registered active ingredients, across all types of products, with more than 2,000 finished (trade-named) products.[775]

Pesticide regulation, especially with regard to the inherent hazards of such products, has always been a priority in Saudi Arabia. In 2009, in response to the death of two children in the country attributed to improper use of a pesticide containing aluminum phosphide, the Ministry of Agriculture elected to ban 23 chemical substances from the market. These 23 substances included bromadiolone, chlorpyrifos, carbaryl, endosulfan, methiocarb, methoxyclor, and others.[776]

While the first regulations specifically addressing the trade in agricultural pesticides began in 1976, legislation on the topic remained largely stagnant until 2006. In 2004, the Supreme Council of the GCC approved "The Regulation of Agricultural Pesticides in the Countries of the Cooperation Council for the Arab Gulf States." In 2006, Saudi Arabia enacted this Regulation into national law in the Kingdom through the issuance of its Executive Regulations.[777] The specific implementation occurred via Royal Decree No. M/67 of 2006, issuing Cabinet Resolution No. 256 of 2006.

Import licenses are issued by the Ministry of Environment, Agriculture and Water for seeds and fertilizers, pesticides, veterinary drugs and vaccines, feeds additives, and biological materials, among others.[778] Interestingly, mandatory labeling requirements for the presence of genetically modified ingredients became effective almost five years earlier, on December 1, 2001.[779] It should be noted, however, that all public health pesticides are registered, legislated, and licensed through the Ministry of Agriculture, and not the Ministry of Health, as might be intuitive.[780]

The registration procedure for agricultural pesticides is fairly involved, and takes about one to two years, including the field trial.[781] Field trials are required for new pesticides that have either not been previously registered or which are being re-registered.[782] As with most other national registration processes, the associated costs will vary greatly depending on the number of crops and pests (label claims) at issue.

The document which begins the process is the "Application Form for the Registration of Middle East and North African Countries."[783] Applicants should pay special attention to the fact that all documents must be presented to the registration authorities in hard copy, *by personal hand-delivery*.[784] In addition, applicants should request from their technical suppliers a "registration certificate of active technical ingredient[s]," which is issued from a variety of governmental authorities, such as the Saudi Arabian Ministry of Agriculture, the Chinese Institute for the Control of Agrochemicals, the Ministry of Agriculture (ICAMA), and/or others. Finally, a certificate of free sale from the manufacturer of the technical grade, duly legalized, should be submitted.[785]

If the substance at issue is a formulated product, then the following items must be provided as part of a complete registration package:[786]

1. A legalized Free Sale Certificate;
2. A legalized Certificate of Registration;
3. A legalized Certificate of Analysis from the producer;
4. An authorization letter from the producer of the finished product to the exporter; and
5. A certificate of storage stability.

The dossier itself should include the following:[787]

1. Technical data information:
   a. Physical & chemical properties;
   b. Impurities data;
   c. Toxicological data;
   d. Eco-toxicological data;
   e. Residue data;
   f. Maximum residue limits;
   g. The method of analysis for technical and residue analysis in soil, water, and air; and
   h. Environmental fate.
2. Finished product data information:
   a. A detailed product composition;
   b. Product specifications;
   c. Intended uses and application rates;
   d. The method of analysis, with the calculation sheet and chromatographs;
   e. Storage stability at both hot and cold test points;
   f. A Material Safety Data Sheet (MSDS); and
   g. A copy of the proposed label.

Similar to the Israeli requirement, the efficiency of the to-be-registered products should be evaluated according to Saudi Arabian climates and conditions.[788]

The Ministry of Environment, Water and Agriculture has its own special accredited laboratories, which make all the requested analyses for the formulated products and residue analyses according to the CIPAC standards.[789] CIPAC is the Collaborative International Pesticides Analytical Council, an international, non-profit-oriented, Non-Governmental Organization (NGO) responsible for promoting the international agreement on methods for the analysis of pesticides and phys/chem test methods for formulations, as well as promoting inter-laboratory programs for the evaluation of test methods.[790]

Related at least in part to pesticide regulations in Saudi Arabia is Ministerial Decision No. 307 of April 21, 2003 (Decision No. 307), officially known as the "Law of Private Laboratories." Article Two of Decision No. 307 directs that the Ministries of Interior, Health, Agriculture and Commerce shall be considered the competent authorities that may issue licenses for private laboratories.[791] Entities wishing to apply for an initial license to operate a private laboratory shall submit the following:[792]

1. A properly completed application;
2. A project study which includes the estimated cost, "technical cadres" (the group of technically qualified individuals to be involved), and "equipments" (to be used);
3. Evidence proving the applicant is a Saudi citizen. If the applicant is a company of "foreign and mixed capital," the company shall have a license from the General Investment Authority which allows them to carry out such an activity;
4. Evidence that the applicant will appoint a Saudi Technical Manager who specializes in "laboratory works," and who will be engaged on a full-time basis; and
5. Evidence that the applicant will provide the necessary support staff ("technical cadres," as above), and the required equipment (again, as above) that "suit the size and nature of the work as determined by the Competent Ministry."

Article Four of Decision No. 307 mandates that the competent department must take a decision on the application within 30 days of the date of submission. If the application is granted, the applicant will be given an initial license authorizing him/her to complete the steps necessary to receive a final license. If rejected, the applicant may appeal the decision before the Competent Minister.

The steps required to potentially receive a final license are detailed in Article Five, and must be completed in not more than six months. These are:[793]

1. The provision of the organizational framework of the laboratory;
2. The appointment of the Saudi Technical Manager (see #4, above) and a submission of an approved copy of his academic qualifications, experience, and training courses;
3. The submission of an approved copy of the academic qualifications, experience, and training courses for the technical cadres;
4. A determination of the installations and equipment needed to operate the laboratory;
5. The Approval Certificate provided by the competent authority previously; and
6. Submission of the approval from the competent authority or municipality which permits the activity in the location.

The applicant must also submit a license fee of Saudi Riyals (SR) 5,000 for the main laboratory, and SR 2,500 for each branch. The license is valid for five years. The initial license will be cancelled if the Article Five requirements are not completed within six months.[794]

One of the more interesting aspects of Decision No. 307 is Article Eleven, the first Article residing under the heading "Assistance by Private Laboratories for the Purpose of Customs Clearance." Article Eleven empowers the Competent Minister, when needed, to issue a resolution engaging private laboratories in the testing of imported commodities for the purpose of customs clearance. Such commodities, subject to a resolution, are defined individually by the Ministries identified in Article Two:[795]

- Minister of Interior: Explosives, chemical materials used to manufacture explosives;
- Minister of Agriculture and Waters (*Note*: in Article Two, the competent authority is named the "Ministry of Agriculture." Here, in Article Eleven, the competent authority is named the "Minister of Agriculture and Waters." As discussed previously in the text, the official name of the Ministry is the Ministry of Environment, Water & Agriculture): All types of living meat animals, all types of ornamental birds and fish, veterinary vaccines, ornamental plants, seeds, and a variety of other items;
- Minister of Industry and Electricity (*Note:* there is a disconnect here with respect to the naming of the Minister/Ministry similar to the foregoing. Article Two names the competent authority as "Ministry of Industry." Here, in Article Eleven, the competent authority is named "Minister of Industry and Electricity." The official name of the Ministry is Ministry of Commerce and Investment): Chemicals (other than dangerous explosives and those requiring a license from the Ministry of Health) imported by national factories; and
- Minister of Commerce (note the difference in names here as well): "All other imported commodities which were not mentioned in the previous paragraphs of this Article."

Article Fifteen details the specific parameters and conditions that the private laboratory must follow under the resolution(s) issued, such as maintaining an office in the customs area for testing, analytical records retention for a period of not less than five years, administrative requirements (e.g. "Putting the license, the organizational framework, the technical sections, the cost of tests and the certificate of approval issued by the authority in a prominent place [at] the entrance of the laboratory"), and maintaining data confidentiality, among other requirements.[796]

The remaining Articles of Decision No. 307 refer to the more procedural aspects of how such samples shall be taken (Articles Seventeen, Eighteen, and Nineteen), the required delivery/holding time (Articles Twenty and Twenty-One), the administrative and documentation aspects relating to the sample/sampling (Articles Twenty-Two and Twenty-Three), and laboratory reporting time (Articles Twenty-Four, Twenty-Five, and Twenty-Six). Article Twenty-Seven lists the ten "Main Sectors" into which the laboratories are classified, Articles Twenty-Eight through Thirty-Three address the "supervision" aspects of such testing, and Articles Thirty-Four through Thirty-Eight present the "General Provisions" – essentially a "catch-all" section for items not addressed elsewhere in Decision No. 307.

Occupational Safety and Health Regulations

Saudi Arabian worker and workplace Occupational Safety and Health (OSH) regulation is particularly interesting, as responsibilities for the various aspects are split between the Ministry of Labor and Social Development in some cases, and the specific employer in other cases. For example, some employer responsibilities with respect to OSH may be:[797]

1. To maintain the worksite in a clean and hygienic condition, provide adequate lighting, and supply drinking and washing water;
2. To take the necessary precautions to protect the workers against hazards and occupational diseases;
3. To post in a prominent place in the workplace the instructions related to work and workers safety in Arabic, in addition to any other language understood by the workers;
4. To train the workers on using safety tools;
5. To appoint a supervisor to educate the workers on OSH procedures, make regular inspections to ensure the safety of equipment, and supervise the performance of OSH rules;
6. To inform the worker, prior to engaging in the work, of the hazards of his job and **require** [emphasis the author's] him to use the prescribed protective equipment; and
7. To supply the workers with the appropriate personal gear and train them on their use.

"Safety Supervisors" at the place of employment have the obligation(s) to:[798]

1. Conduct regular inspection(s) of the workplace;
2. Investigate work injuries;
3. Supervise the procurement of appropriate Personal Protective Equipment (PPE);
4. Educate workers on preventative measures;
5. Prepare accurate statics on work injuries; and
6. Supervise the implementation of OSH programs.

The main regulations under which OSH issues in Saudi Arabia are generally addressed are the Labor Law (also known as Royal Decree No. 21 (6/9/1389 H) on Safety in the Workplace, promulgated on November 15, 1969) and the Labor and Workmen Law. The Labor Law was most recently amended through Ministerial Decision No. 1982 of April 6, 2016. The new implementing regulations became enforceable immediately upon their publication in the April 22, 2016 Official Gazette, and superseded the previous Ministerial Resolutions No. 1/693 and 1998/1.[799]

The Ministry of Labor and Social Development holds responsibility for the "development and use of the Kingdom's human resources. It is responsible for manpower planning, labor relations and the general monitoring of all matters relating to employment affairs… [as well as] labor disputes, employment in the private sector and labor visas."[800] The Labor Law was one of the first OSH regulations promulgated in the country.

While some exceptions are laid out in Articles 3 and 4, Article 2 of the Labor Law clearly defines those who are subject to regulation – and fairly broadly – as "Any contract under which any person undertakes to work for the account of an employer under the latter's direction or control in consideration of a wage."[801] Buried in the Labor Law, in Chapter 1 (*Note:* in the language of the text, there is a "Chapter One," which is the first chapter in the whole of the Labor Law, and also a "Chapter 1," which occurs almost at the end of the Law), under the heading "General Organization of the Labor Inspection Apparatus," Article 6 provides the specific mandate for worker safety and health protection. Article 6 requires that the central section for labor inspection shall assume the following duties: "Looking after the safety and health of the workmen. Protecting them against the hazards of machinery, occupational diseases and work injuries, and promoting hygienic and preventive consciousness by all means possible."[802] Thus, at least in terms of the legislative text, the Ministry of Labor has responsibility for actions that would, under other national laws elsewhere, fall to the private entity or operation.

Many of the remaining sections of the Labor Law are largely administrative in their content, or relate to areas outside OSH regulation. While even the 2016 amendments to the Labor Law contained very little that was applicable to workplace OSH, with respect to chemical substances, it did establish "the process through which the Labor Office must be notified of work injuries. It also explains that the Labor Office will determine the compensation that is owed to the worker based on a medical report specifying the degree of invalidity."[803] Finally, the New Implementing Regulations specify the minimum content to be contained in medical first aid kits.[804]

Waste Regulations

Saudi Arabia has multiple regulations in place to manage waste throughout the Kingdom, from generation to storage, transportation, and disposal.

The Standard on Waste Transportation, in force since March 24, 2012, has the manifest objective of establishing the "requirements necessary to ensure that the transportation of waste in KSA [Kingdom of Saudi Arabia] is undertaken in a responsible manner to safeguard the protection of the environment and the community against potential accident, spills and pollution."[805] The Standard "incorporates the international requirements for dangerous goods and substances in relation to the transportation of hazardous waste to ensure consistency is achieved."[806] In addition, it endeavors to reduce road accidents involving waste transporters, to provide waste transporters with a consistent system regarding waste classification and waste labeling, to facilitate transboundary waste movement, as well as to simplify waste transportation processes to ensure easier compliance.[807] All types of waste generators, be they hazardous, non-hazardous, and/or inert, must comply with the Standard.

Safety Data Sheets and Labels

Labeling and marking requirements are compulsory for any products exported to Saudi Arabia… The Ministry of Commerce and Industry implements SASO guidelines through its inspection and test laboratories at ports of entry.[808]

Labeling is particularly important for companies marketing food products, personal care products, health care products, and pharmaceuticals. SASO has specific requirements for identifying marks and labels for various imported items.[809]

SASO establishes the standardized specifications for chemical packaging in terms of their type, size, color, and the signs and instructions that should be labeled on them.[810] Furthermore, "containers holding hazardous materials must be individually labeled to reflect the actual contents of the container, and the label should include the contents and associated hazards or a unique identification that is cross-referenced to a document which lists the contents and hazards."[811]

Labeling of chemical substances is additionally managed under Section 5 of the Law of Chemicals Import and Management. Section 5 directs that SASO "shall set standard specifications for containers of chemicals, in terms of type, size, [and] color as well as mandatory labels and instructions."[812]

Saudi Arabia has not implemented GHS yet. However, an SDS is required for all hazardous materials, although the Implementing Regulation delegates the authority for defining "hazardous substances" to the local level.[813]

9

United Arab Emirates

National Overview

The Trucial States of the Persian Gulf coast initially consisted of eight states: Abu Dhabi, Ajman, Al Fujayrah, Ash Shariqah (Sharjah), Dubayy (Dubai), Kalba, Ras al-Khaimah, and Umm al Qaywayn. On December 2, 1971, six of these states – all except Kalba and Ras al-Khaimah – merged to form the United Arab Emirates (UAE), with Abu Dhabi as the capital. Abu Dhabi, located in the largest and wealthiest of the six emirates, "has the vast majority of oil and gas reserves in the UAE, [and] has made significant investments in establishing aerospace, nuclear power, defense, information technology (micro-processing), petrochemical and clean-tech industries."[814] The UAE is relatively small, with a land area of just over 83,000 sq. km.[815] The population is heavily concentrated to the northeast, and the three largest emirates – Abu Dhabi, Dubai, and Sharjah – are home to nearly 85% of the population.[816] The UAE is the Middle East's second largest economy, after Saudi Arabia, and one of the wealthiest countries in the region on a per capita basis.[817]

Governmental Structure

The UAE is a federation of monarchies, governed by a mixed legal system of Islamic law and civil law. Similar to many other countries examined, the governmental structure of the UAE is organized into three distinct branches: Executive, Legislative, and Judicial. The executive branch is led by the Chief of State (President), and also consists of the Vice-President and Prime Minister (Head of Government), two Deputy Prime Ministers, and a Cabinet (Council of Ministers) announced by the Prime Minister and approved by the President.[818] The President and Vice-President are indirectly elected by the Federal Supreme Council, for a five-year term, with no term limits.[819] The Prime Minister and the Deputy Prime Minister are appointed by the President. The Federal Supreme Council is comprised of the rulers of the six emirates discussed earlier, establishes general policies, sanctions Federal legislation, and is the highest constitutional authority in the UAE.[820]

The legislative branch is comprised of the Federal National Council (FNC, *Majlis al-Ittihad al-Watani*), which contains 40 seats. 20 of the seats are indirectly elected by an "electoral college," and the remaining 20 are appointed by the rulers of the six emirates.[821] Perhaps interestingly, the FNC elects its members in an electoral college format,

Chemical Regulation in the Middle East, First Edition. Michael S. Wenk.
© 2018 John Wiley & Sons Ltd. Published 2018 by John Wiley & Sons Ltd.

similar to how the United States elects its President, although the college is comprised of significantly more electors (224,279 at the close of 2015, versus 538 in the United States).[822] The UAE has a two-tier system: Federal law, which applies to all six emirates, and local laws, which are confined to the emirate in which they are enacted.[823]

Finally, the judicial branch consists of a Supreme Court comprised of a Court President and four judges, with jurisdiction limited to Federal cases.[824] These judges are appointed by the President, following approval from the Federal Supreme Council, and serve until retirement or the expiration of their appointed terms.[825]

The UAE publishes its Official Gazette on a monthly basis, and includes laws and decrees issued. The Gazette may be accessed at https://www.ecouncil.ae/en/Official-Gazette/Pages/default.aspx.

Key Chemical Regulatory Agencies

> Throughout the history of the United Arab Emirates, the environment has been a priority concern with strong institutional mechanisms to enhance environmental protection – starting with the foundation of the Supreme Committee for Environment in 1975, the Federal Authority for Environment in 1993, and the Ministry of Environment and Water in 2006. These developments reflect the increasing concern placed on environmental issues, at both the local and global levels, and the importance of addressing these concerns through the development of appropriate policies and regulations.[826]

The UAE's Ministry of Environment and Water (MOEW) was initially the central authority for chemical substance management. Established in February 2006, the MOEW arose from a national vision to strengthen the commitment of the UAE to the environment and sustainable development.[827] The MOEW replaced the former Ministry of Agriculture and Fisheries, assuming all its powers and functions.[828] Additionally, the Federal Environment Agency and the General Secretariat of Municipalities had their functions and authorities transferred, to be accommodated within the MOEW's mandate.[829] Within the MOEW, there are five key sectors or divisions: Environmental Affairs, Agricultural & Animal Affairs, Water Resources & Nature Conservation, Support Services, and Regions Sector.

In 2016, the UAE expanded the role of the MOEW to manage all aspects related to international and domestic climate change affairs.[830] As such, the MOEW was renamed the Ministry of Climate Change and Environment (MOCCAE). The "Ministry's Goals" and the "Mandate of the Ministry," both available on the MOCCAE web page, reveal that the MOCCAE has roles and responsibilities beyond "merely" protecting the natural environment, as will be seen later in this chapter.

Key Chemical Substance Regulations

The UAE has a strong interest in chemical substance management, with specific attention given to hazardous chemicals administration. As such, it works with a variety of regional bodies, such as the GCC and others, to develop standards and specifications, as well as regional and international chemicals agreements such as the Montreal Protocol on Substances that Deplete the Ozone Layer, the Stockholm Convention on Persistent Organic Pollutants, the UN Framework Convention on Climate Change, and the Basel Convention on the Control of Transboundary Movements of Hazardous Wastes and their Disposal, among others.

In 1999, Federal Law No. 24 ("For the Protection and Development of the Environment") was promulgated under the oversight of the Federal Environmental Agency. Federal Law No. 24, which has been deemed by some as the original chemical control law in the UAE, contains the following goals:[831]

1. Protection and conservation of the quality and natural balance of the environment;
2. Control of all forms of pollution and avoidance of any immediate or long-term harmful effects resulting from economic, agricultural, industrial, development, or other programs…;
3. Development of natural resources and conservation of biological diversity…;
4. Protection of society, human health and the health of other living creatures from activities and acts, [sic] which are environmentally harmful…;
5. Protection of the State environment from the harmful effects of activities undertaken outside the region of the State; and
6. Compliance with international and regional conventions ratified or approved by the State regarding environmental protection, control of pollution and conservation of natural resources.

Chapter 1 of Federal Law No. 24, "Development and the Environment," addresses the setting of "standards, specifications, principles and regulations for the assessment of environmental impact of projects and establishments," and details in Article 3 the steps to be taken as part of the application process. These steps involve identifying categories of projects which, as part of their inherent nature, may cause harm to the environment, areas of special environmental importance (e.g. wetlands, coral reefs), and natural resources and major environmental problems of special importance.[832] Article 4 mandates the Federal Environmental Agency, and others as appropriate, to evaluate the environmental impact(s) of projects and establishments, and to take a decision within one month from the date the application is submitted.[833] Such entities must have a license granted before beginning operations, with the requirements for a complete application detailed in Article 5. Those entities granted a license are required to regularly monitor and analyze wastes produced, and to send reports regarding the same to the Agency and the competent authorities.[834]

Section 2 of Federal Law No. 24 is somewhat unique, in that it relates specifically to the issue of sustainability in the Kingdom, a topic which is not widely found in other Middle Eastern country regulations. Article 9 of the "Environment and Sustainable Development" section mandates "[a]ll Concerned Parties… shall consider aspects of protection of the environment, control of pollution and rational use of natural resources when developing economic and social plans and when establishing and executing of [sic] projects."[835]

Of further interest with respect to the emphasis that Federal Law No. 24, and thereby the Kingdom, places on sustainability is Article 10, which expressly directs "[t]he balance between technological capabilities available and economic cost shall be considered when determining such measurements and standards [for environmental protection] without undermining the requirements for the protection of the environment and control of pollution."[836] Additionally to this point, Article 12 of Section 2 specifically sets out a prohibition on the hunting, killing, or capturing of "birds, wild and marine animals… [or] to damage birds' nests or destroy their eggs," presumably as a means of wildlife protection.[837]

The remaining chapters of Federal Law No. 24 discuss a variety of environmental protection measures, most again related to sustainability, and the means by which they are to be accomplished. For example, Section 1 of Chapter 2 addresses "Protection of Water Environment," including coasts, beaches, and seaports, the living and non-living natural resources of the marine environment, and the protection of drinking and groundwater. Section 2 establishes a prohibition, in Article 21, of "discharging or disposing of oil or oil mixture into the marine environment," and in Article 23 for responsibility for the remediation of at-sea collisions. With respect to such collisions, irrespective of whether they are deliberate [it is unclear to the author why there would be a "deliberate" collision at sea, but the language of the Article expressly includes the term] or the result of negligence, the "owner and transporter [of the vessel(s) will be] jointly responsible for the payment of all costs of damages, compensation and control incurred."[838] Article 26 directs that any "marine means" transporting oil in the Kingdom "shall be equipped with the necessary equipment to undertake combating operations during the occurrence of pollution from the same marine means," and Article 27 specifically prohibits the discharge of "hazardous substances" or "harmful substances or wastes" directly or indirectly into the marine environment.[839]

In 2001, Cabinet Resolution/Decision No. (37) "Regulating Federal Law No. (24) of 1999 Concerning Protection and Development of the Environment" (Decree No. 37) was promulgated into national law. Decree No. 37 defined "hazardous substance[s]" as "solid, liquid or gaseous substances having properties that are harmful to human health or have adverse impact on the environment, such as toxic substances, explosive, flammable, or ionizing radioactive substances" and "harmful substance[s]" as "all substances such as chemicals, biological or radioactive materials, leading to harmful effects on human health or the environment directly or indirectly."[840]

In mid-2016, enforcement of Ministerial Decree No. 783 of 2015 began, requiring companies producing, using, importing, or exporting hazardous chemicals to ensure that they do not use any of the banned chemical substances listed therein.[841] Additionally, companies must not import the restricted chemicals listed in the same Decree before obtaining the approval of the MOEW.[842]

Pesticide Regulations

Pesticides should be registered with the MOCCAE before their import or production in(to) the UAE. An import permit is required.[843]

On September 28, 1992, the Ministry of Agriculture and Fisheries enacted "Federal Law Number (41) of the Year 1992 Concerning Pesticides" (Federal Law No. 41), which addressed pesticides used in agricultural applications. Federal Law No. 41 established a definition of "pesticide" which was simultaneously consistent with how a variety of other global regulations define the term, while also exceptionally broad. According to Federal Law No. 41, a "pesticide" is:

> Any material or mixture of materials manufactured for protection or controlling plants against any kind of pests. This shall include insect vectors that transmit diseases to man and animal, undesired plants and animals which cause damage, or by any means interfere during the stage food production, agricultural production, wood, wooden crafts or forage, and fodders during their manufacturing, storing, transporting and marketing, swell to materials that the animals take to control pests, insects, mites or other pests, weather internal or external insects, or parasites.[844]

Further:

> The above shall include the materials used for organizing the growth of plants, downfall of leaves, or drying the plants, turning dawn fruit's trees, protection of fruit fading before harvesting, and the materials, which are used for protecting crops against either before or after harvesting, in order to protect the crop against degeneration during transmitting or storing.[845]

Article 2 of Federal Law No. 41 directs that the provisions of the law shall be applicable to the following:[846]

A - All kinds of pesticides;
B - Ingredient material: The active part of pesticides;
G[sic] - Commercial formulation of pesticide. The final and active formulation of the pesticide; and
D - Additives: If separately sold to be used with the pesticide, i.e. the additive materials which are added to the formula of the pesticides in order to improve its activity.

Pesticides that are imported by universities and research centers for scientific purposes, on the condition that such parties take all safety precautions and ensure that the pesticides are not circulated in the country, are specifically exempt from the scope of the Law. Additionally exempt are pesticides that are imported into the UAE for the purposes of re-exportation.

Article 3 of Federal Law No. 41 expressed the legal directive that permission from the competent authority (registration) is required, while Article 4 established the Committee of Agricultural Pesticides, charged with "determin[ing] [the] kinds of pesticides allowed for circulation, their specifications [and] the procedures of their registration."[847] Other significant Articles of Federal Law No. 41 are:

- Article 6, which directs the Minister to issue decisions "especially which [sic] concerned with the following": pesticides which are prohibited from circulation or import due to their dangerous effects on health and the environment, and those which should only be used by "specialists." The Article also sets forth the directive to issue "terms and conditions of licensing," as well as the "specifications, data of pots of pesticides, stickers [labels and related items] and illustrated directions" and for sampling and analysis of pesticides.

- Article 8, which gives those who have been selected as "judicial detection inspectors" the authority to enter places where such pesticide activity takes place (except for residential areas) to ensure "implementing its [the Law's] provisions and the decisions issued in execution thereof."
- Article 9, setting out the penalties for violation of Federal Law No. 41. These include imprisonment for a period of not more than six months, and a fine of not less than 20,000 dirhams, and not to exceed 100,000 dirhams. If the violation caused harm to the health, or led to the death, of a human, then the appropriate terms of Federal Law No. 3 of the year 1987 "concerning the issuance of the punitive law" shall also be applied.

The applicable base service fees from the Ministry are as follows:

1. Pesticides import permit (active ingredient)[848]
   - AED 2000 to request pesticide sample testing or re-testing;
   - AED 500 to request an import permit for consignment or samples of pesticides; and
   - AED 500 to request release of a consignment or pesticide sample or active substance.
2. Pesticides registration certificate[849]
   - AED 3000 per request to issue or renew a registration certificate for a pesticide.
3. Release of pesticides consignment (active ingredient)[850]
   - AED 2000 per request for pesticide sample testing or re-testing; and
   - If non-compliance is found with respect to the import regulations, a fine of ten times the prescribed fee for the import permit shall be imposed and required.

The data points and applicable registration steps required to compliantly register a pesticide in the UAE are as follows:[851]

1. A pesticide registration certificate from the country of origin and the pesticide registration certificate from the member countries of the Economic Cooperation Organization (for chemical and organic pesticides) and development. [*Author's note:* this is the verbatim language from the MOCCAE website. However, it is believed that the organization being referred to is actually the OECD];
2. A valid organic product certificate, for organic pesticides, issued from a certifying body accrediting Euro, American, or Japanese organic production standards;
3. A certificate from the exporter or producing company showing that the material does not contain any genetically modified materials, and no materials produced from genetically modified organisms are used in its production;
4. A pesticide composition certificate proving that its components conform in quality and quantity to the materials produced, issued, and certified in the country of origin. In addition, an analysis certificate of the active material, showing purity rate and impurities;
5. A local pesticide package label copy showing, in both Arabic and English languages, all data in conformity with the label date of the country of origin. Additionally, a copy of the package label from the country of origin, in English, and translated legally into Arabic;

6. A local pesticide package label copy showing, in both Arabic and English languages, all data in conformity with the label in its country of origin on the registered pesticide (for pesticides (organic)) and shall abide by putting the Emirates trade mark (organic); and

7. An authorization letter for the facility from the producing company authorizing it to register and market the pesticide.

The UAE, through the MOCCAE, has undertaken a fairly extensive IT effort to make e-filing possible for a variety of aspects, and to reduce the administrative burden when doing so. As an aside, to this end, the MOCCAE has implemented a "Customer Happiness Charter" on its website, noting therein: "We are committed to providing the most advanced services in line with the highest applicable standards. As part of this priority, we aim to exceed our customers' expectations and ensure positive encounters and happiness" and established "Customer Happiness Centers" throughout the country.[852,853]

Among these e-filing opportunities are pesticide and agricultural/fertilizer registration applications. Further to the extensive IT resources invested by the MOCCAE with respect to registration, they have posted a YouTube video describing how to register for the service, which may be found at https://youtu.be/HpupN9W9dcY. In addition, it has published a smartphone app for the same purpose. Interested parties may search "MOCCAE" in their respective app stores, and then click "search."

To access the registration web portal, applicants should navigate to the MOCCAE homepage, and then select "Services for Businesses" at the bottom of the page. From the ensuing page, applicants may register for a variety of programs/uses, such as the "Fertilizers and Agricultural Conditioners Package," "Aquaculture Farms Permits Package," "Hazardous Waste Package," and "Pesticides Package." A "package" contains links to the data set necessary for a complete application submission. As an example, selecting "Pesticides Package" and then "Pesticides import permit (active ingredient)" on the ensuing screen will navigate the user to a dedicated page for such registration. He or she may then click "Start the service," which will step through the electronic registration process, "form download," which will provide the applicable forms for the process, or "user manual," which will provide an overview of the process.

The MOCCAE has established an electronic directory of registered pesticides, downloadable as a Microsoft Excel file from https://www.moccae.gov.ae/en/knowledge-and-statistics/pesticides.aspx.[854] The "Registered Pesticides List" is searchable and sortable by tradename, active ingredient, "local company," "producing company," and other relevant fields.

On December 20, 2010, the MOEW enacted "Ministerial Decision No. (849) for the Year 2010 on the Amendment of the Ministerial Decision No. (554) for the Year 2009 Concerning the Prohibited and Restricted Use [of] Pesticides in the United Arab Emirates" (Decision No. 849). Article 1 of Decision No. 849 directed that

> individuals, companies, establishment [sic] and the public and private sectors shall be banned from production, manufacture, formulation circulation [sic], import and use of any of the following varieties of pesticides shown… in the lists numbers [sic] (1, 3, 5) and attached to this decision.[855]

Returning to the sustainability discussion, Article 2 of Decision No. 849 mandates that "companies, establishment [sic] and public and private sectors" which have any of the pesticides contained in lists 1, 3, and 5 must notify the competent authority and concerned bodies of a plan to "eliminate these pesticides within 30 days from the issuance date of this decision."[856] Careful readers will note that "individuals" are not included in Article 2's action, although they are included in Article 1. The exact reason for this is not known. It may be a translation issue from the Arabic to English text, as no other direction or explanation for this is given in the text of the Decision.

In 2015, the MOCCAE issued "Ministerial Decree No. 799 for 2015 Introducing Amendments to Ministerial Decree No. 236 for 2014 Governing the Registration and Import of Pesticides" (Decree No. 799). As per Decree No. 799, the distribution of any pesticide without proper registration from the MOCCAE is expressly prohibited. Additionally, pesticides which are banned may not be registered or imported into the state, and companies are barred from using pesticides without the MOCCAE's approval.[857]

Decree No. 799 also forbids "using similar brand names for registered pesticides or labeling the product after its active ingredient."[858] Once the MOCCAE grants its approval for the importation of a pesticide, it will issue a certificate of registration, valid for five years or until the expiration of the certificate issued by the country of origin.[859] The certificate of registration, however, may not be transferred or otherwise assigned.[860] Decree No. 799 further states "an import permit issued to a facility is non-transferrable and… imported pesticides must be imported from the manufacturing company in the country of origin or its branches."[861]

The MOCCAE has communicated that it may repeal or reject an application for registration if there is incomplete or incorrect information and/or documents, if the authority in the country of origin of the product has determined that it will not renew the product's registration, if the product is shown to not be efficacious, and/or if it is found to be harmful to the environment, human health, animals, or plants.[862] Should a registration be cancelled, or the product banned from the UAE, the company is required to withdraw the pesticide from the market and export any stocks from the UAE "within 90 days after the date of issuance of the ban order or cancellation of the product registration."[863]

Registrants should ensure that the local label of the registered product matches the data provided on the label issued by the pesticide product's country of origin.[864] In addition, import permits issued to a facility are non-transferrable, and pesticides which are being imported into the UAE must be imported from the manufacturing company in the country of origin, or its branches (e.g. not from a third-party warehouse).[865]

In June 2016, the MOCCAE issued several new measures respecting the registration and importation of various types of pesticides. Ministerial Decree No. 464 of 2016, amending Ministerial Decree No. 766 of 2015, updated the procedures for registering and importing pesticides in the UAE.[866] Compliance with these new measures is mandatory for UAE pesticide companies. The regulations were passed to prevent the import of pesticides that could harm consumers and the environment.[867] Among the measures was a provision that would allow companies to handle and use pesticides whose registration has expired for a period of 180 days after their expiration date.[868] In addition, the updated procedures allow companies that intend to register pesticides with the MOEW "to submit to the Ministry the trade certificates of such pesticides of the country of their origin instead of their registration certificates when the registration certificates are not available."[869]

Occupational Safety and Health Regulations

Within the UAE, there is no specific OSH legislation in place. Indeed, "[t]here is no statutory body in the UAE to oversee EHS and ensure appropriate PPE is used."[870]

> There are various laws that address aspects of health and safety, albeit in general terms. For example, Ministerial decision 32 [sic], enacted in 1982, outlined provisions in the labor law for the construction industry, but it did not provide technical requirements or standards that could be used to assess whether an entity is in compliance. Similarly, while the penal code deals with acts or omissions causing personal injury or death, the provisions are not specific to safety and health or construction. They equally could apply to someone causing a death by driving dangerously.[871]

Worker and workplace safety is primarily managed under Federal Law No. 8 of 1980, known officially as the "UAE Labor Law and its Amendments" and informally as the "Federal Labor Law." The Labor Law was updated most recently in 2016. "In addition, Order No. 32 of 1982 on Protection from Hazards and Ministerial Decision No. 37/2 of 1982 are also fundamental federal laws. Besides that, other regulations address specific hazards on a national level."[872] The Minister of Labor and Social Affairs has responsibility for OSH in the UAE, in consultation with the Ministry of Health.[873]

Federal Law No. 8 is highly protective of UAE nationals in its scope, with respect to the areas in which non-citizens may work, and the processes by which they may apply for and be approved to do so. Article 9 of Chapter II specifically recognizes that "[w]ork is a right of the United Arab Emirates Nationals. Others may not be employed in the United Arab Emirates except as provided for in this Law and its executive orders" and Article 11 of the same chapter directs that "[a] Section at the Labor Department shall be created for the employment of nationals."[874] As such, employment in the UAE may be viewed as a fundamental human right.

As Federal Law No. 8 is fairly broad in its scope – in that it encompasses a wide variety of provisions (e.g. employment of minors, holidays and leave, remuneration, and so forth) and occupational jurisdictions – the following sections will highlight those Articles which are of particular interest to workers in the chemical industry in the UAE:

- Article 3 of Chapter 1 specifically exempts from its purview "officials, employees and workers" of a variety of governmental entities (e.g. Federal government, municipalities, local public departments, and so forth), members of the Armed Forces of Police and Security, domestic servants, and those employed in agriculture or pastures (except for those engaged in processing such products or operating or repairing machines required for such).[875]
- Articles 4 and 5 of Chapter 1 address the burden placed upon the employer in situations where worker(s) file a case against them under Federal Law No. 8. As per Article 4, "all amounts payable to the employee or his beneficiaries under this Law shall have lien on all the employer's movable and immovable properties" and as per Article 5, "[c]ases filed by employees or their beneficiaries under this Law shall be

exempted from court fees at all stages of litigation and execution and shall be expeditiously heard." As such, the Law appears to tilt in favor of the worker when a claim is brought against the employer.

- While women are generally permitted to work in the UAE, except at "night" (defined in Section 3, Article 27 to be "a period of not less than eleven consecutive hours including the period from 10 p.m. to 7 a.m."), Article 29 of the same Section directs that women "may not be employed where jobs are hazardous, harmful or detrimental to health or morals..."[876] As such, a variety of roles in chemical manufacturing and related industries would appear to *de facto* exclude women.

- Section 3 of Article 53 discusses the data which employers are required to keep on their employees, including "injuries and vocational diseases sustained by him."[877] Further, Article 54 mandates that an employer with 15 or more employees "shall keep... [a] Register of work injuries [containing] [a]ll work injuries accidents [sic] and vocational diseases sustained by the employees... recorded in this register as soon as they are brought to the knowledge of the employer."[878]

- Chapter IV, Article 65, addresses that the "maximum normal working hours for adult employees shall be eight hours per day or forty-eight hours per week."[879] However, "working hours per day in respect of hazardous work or work detrimental to health, may be decreased by decision of the Minister of Labor and Social Affairs."[880]

Chapter V of Federal Law No. 8 is expressly entitled "Safety, Protection and the Health and Social Care of the Employees," and contains the following sections:

- Article 91 refers directly to the use of PPE, noting:

> Every employer must provide adequate means of protection for the employee from the hazards of injuries and vocational diseases that may occur during work as well as the hazards of fire and other hazards arising from use of machines and other tools, and he must apply all other means of protection as approved by the Ministry of Labor & Social Affairs.[881]

In addition, the employee bears specific responsibility for his safety as well:

> [A]nd the employee must use protective equipment and clothing provided to him for such purpose and he must abide by all instructions of the employer aiming at his protection from dangers and must not act in a way that may obstruct the application of said instruction.[882]

- Article 92 requires employers to display "at a conspicuous point in the place of business detailed instructions concerning methods to prevent fire and protect employees from dangers while they perform their duties";[883]

- Article 93 mandates the provision by employers of a "medical aid box" and to have it be used "by a specialist in handling first aids," with a minimum of one box per 100 employees;

- Article 94 addresses the requirement for "proper cleanliness and ventilation" in the workplace, and to ensure such aspects as "adequate illumination, potable water and toilets";

- As per the requirements of Article 95, employers are obligated to appoint a physician who must conduct a full medical check-up at least once every six months "regularly for his employees who are exposed to the danger of infection with any of the occupational diseases reserved the schedule attached hereto."[884] The results of the examinations must be recorded in the employees' personal files and, in the case of occupational diseases, must be reported "instantly" to the Labor Department, once their presence is confirmed. Relatedly, Article 96 mandates that employers must provide their employees with a means of medical care, although it does not specify whether such care needs to be provided at the work location or elsewhere.
- Strangely, Articles 97 and 98 are identical in their text. They both require the employer or his representative to keep employees informed of the dangers involved with their work, and of the measures the employees must take to prevent them. Article 100 prohibits the employee acting "in any way that may contravene enforcement" of the foregoing instructions, and/or the methods placed for safety and health protection.[885]

Waste Regulations

On April 27, 2017, "Regulation No. 10, 2017, Restrictions on the Use of Hazardous Materials in Electronic and Electrical Devices Control Scheme" (RoHS regulation) was published in the official journal, coming into effect the following day, under the purview of the Emirates Authority for Standardization and Metrology (ESMA). The RoHS regulation is based largely, but not completely, on the initial EU's RoHS Directive (Directive 2002/95/EC). Both regulations encompass those entities which were selling Electrical and Electronic Equipment (EEE). The UAE's RoHS regulation established three specific deadlines for EEE firms, as follows:[886]

1. By January 1, 2018, restrictions on lead, mercury, cadmium, hexavalent chromium, polybrominated biphenyls (PBB), and polybrominated diphenyl ethers (PBDE) will enter into force. The exception to these restrictions would be equipment listed in Category 11 of the RoHS regulation, "Other EEE";
2. By January 1, 2020, restrictions on lead, mercury, cadmium, hexavalent chromium, PBB and PBDE in medical devices, in-vitro diagnostic medical devices, monitoring and control instruments, and industrial monitoring and control instruments will enter into law. In addition, limitations on di(2-ethylhexyl) phthlate (DEHP), benzyl-butylphthalate (BBP), dibutyl phthalate (DBP), and diisobutyl phthalate (DIBP) in EEE would apply, with respect to cables, spare parts for repair, reuse, updating of functionalities, or upgrading the capacity of EEE; and
3. Finally, by January 1, 2022, restrictions on lead, mercury, cadmium, hexavalent chromium, PBBs and PBDEs in cables and spare parts of medical devices, in-vitro diagnostic medical devices, monitoring and control instruments, and industrial monitoring and control instruments will come into national law. Additionally, there would be in place restrictions on DEHP, BBP, DBP, and DIBP in medical devices, in-vitro diagnostic medical devices, monitoring and control instruments, and industrial monitoring and control instruments, including cables and spare parts.[887]

Article 5 of the RoHS regulation details the four specific elements which are required for compliance:[888]

1. Compliance of the specific products and/or product types must be verified by an entity that has been approved by ESMA;
2. The requirements of the UAE's Compliance Evaluation System, as detailed in Form A of the RoHS regulation, must be demonstrated to have been met;
3. The technical specifications, requirements, terms, and conditions which are outlined in the RoHS regulation must be objectively met; and
4. "Obtaining information about the design, specifications, testing and inspections reports and any other relevant documentation about the compliance of the product by the product manufacturer, as deemed relevant" by ESMA.[889]

Safety Data Sheets and Labels

The UAE requirements for SDS may be found in Decree No. 37. While the information to be provided does not generally correspond to that set out in other global SDS standards, documents should include the following data:[890]

- Scientific and commercial name of the hazardous substances, and its chemical composition;
- UN number and CAS number;
- Physical and chemical properties;
- Risk level of the substance and its health and environmental impact;
- Volume, date and time of expected transfer;
- Importation purpose;
- Optimal storage and disposal options;
- Measures to be taken in case of accidental spillage;
- Measures to be taken in case of fire;
- Certificate of origin and inspection from the exporting country;
- Production and expiration dates;
- First aid in case of injuries from eye or skin contact, inhalation or ingestion; and
- PPE and clothing.

Within the UAE, goods such as toys, food, chemical products, industrial products, drinking water, textiles, and cosmetics have product-specific labeling requirements.[891] The labeling requirements for chemicals are codified in Cabinet Resolution No. 12 (2007), specifically in Article 28. A compliant product label will contain the following items:[892]

- The kind, nature, and components of the goods;
- The name of the goods;
- The date of production or packaging;
- Expiry date;
- Net weight;
- Country of origin (to include the phrase "made in" before the name of the country, and not to put a flag of another country other than the country of origin);
- Country of export (if any);
- How to use, if possible; and
- The appropriate unit of measure and weight for the goods.

Abu Dhabi

Similar to how the Saudi Arabian cities of Jubail and Yanbu presently have their own regulatory systems in place to manage chemicals, Abu Dhabi maintains its own set of regulations on the topic. The city is the second most populous city of the UAE, as well as being the capital of the Emirate of Abu Dhabi, the largest of the UAE's six emirates. As such, Abu Dhabi has UAE regulations to abide by, but has implemented additional ones specific to the city/Emirate.

In addition to legislation, Abu Dhabi employs Standard Operating Procedures (SOPs) to manage chemicals. While these do not always have the force of law, they are recognized to be a common and acceptable way to achieve a variety of results. For example, Tadweer Waste Treatments LLC, a Dubai-based company whose mission is "to create a quality environment through dedicated leadership and educational partnerships to reduce wastage while promoting recycling and resource conservation," partnered with the Center of Waste Management in Abu Dhabi to create the "Standard Operating Procedure for licensing [sic] of Hazardous Waste Service Providers in the Emirate of Abu Dhabi."[893]

Law No. 16 (2005), pertaining to the "Reorganization of the Environment Agency–Abu Dhabi (EAD)," established the EAD as the competent authority in the Emirate of Abu Dhabi for environmental protection.[894] It has produced a range of SOPs which set out its approach to permitting for commercial, industrial, light industrial, chemical, hazardous materials, and infrastructure projects.

The EAD has established a "Chemical and Hazardous Materials Management System," responsible for regulating and controlling the use and entry of such substances into the Emirate. Among other aspects, the system aims to provide tools for building a cradle-to-grave management system for such substances used in the Emirate. The benefits of this system include public access to relevant federal and local laws and regulations, a retrieval system for MSDS, information for select chemicals, lists of regulated materials (chemicals, pesticides, wastes), and a database of chemical and/or radioactive materials imported into the Emirate.[895]

In 2009, Abu Dhabi enacted Decree No. 42, an innovative law which implemented a comprehensive Environment, Safety and Health Management System (EHSMS) framework. The EHSMS is "intended to be a performance based management system aiming to achieve excellence in the management and protection of [HSE] through partnership between the government and private sectors to ensure that economic activities… are undertaken in a responsible, safe and sustainable manner."[896] However, EHSMS is not meant to replace existing laws relating to the environment and health and safety.

The EHSMS will be implemented using an iterative approach, beginning with specific economic sectors, for which, once approved by the EAD, the sector will have its own regulatory authority.[897] For each sector, certain entities within the sector are to be nominated, and are required to develop their own entity EHSMS. "Each entity must then actively audit their own compliance with their EHSMS and must undergo annual third party [sic] compliance auditing as well. Environment, Safety and Health compliance is to be reported to the relevant sector regulatory authority."[898] Furthermore, any investor "establishing or acquiring a business affected by this legislation will need to ensure his/her organization has or can acquire the management capabilities to embed the requirements of the EHSMS framework into the business's environment, safety and health management systems."[899] Thus, the EHSMS system is neither static in terms of its development nor staid in terms of the entities to which it applies.

In 2012, Abu Dhabi's Environment, Safety and Health Center updated its environmental health, safety and management system, re-branded as AD EHSMS (the "Framework"). "[The] Abu Dhabi Emirate Environment, Safety and Health Management System (AD EHSMS) Framework is an Abu Dhabi Government Initiative that was developed to control environmental impacts resulting from workplaces and to ensure safe and healthy conditions for all workers in the Emirate."[900] Chief among the details of the Framework was the adoption of several mandatory documents which regulated chemicals

> with the aim of harmonizing regulatory requirements across local and emirate levels of government, eliminating duplicative processes and aligning requirements with international standards... The code of practice stipulates that classification of hazardous materials must comply with "applicable international model regulations," meaning that manufacturers may choose from among such regulations including the United Nations Globally Harmonized System of Classification and Labeling of Chemicals (GHS).[901]

Abu Dhabi's "Code of Practice EHS RI CoP 1.0 – Hazardous Materials. V.3 of 2016" indicates that manufacturers "shall assess and classify material by reference to applicable international model regulations (e.g. UN: Global Harmonization System), laboratory analysis, expert judgement, and/or weight of evidence to if materials create a [hazard]." Currently, all revised editions of the UN GHS System are accepted. The EHSMS is meant to complement existing systems. If a contradiction occurs, the local and Federal regulations take precedence.

With respect to OSH regulation, the Abu Dhabi Occupational Safety and Health Center (OSHAD) functions as the competent authority for OSH issues in the Emirate.[902] It supervises the implementation of the Framework by governmental departments, sectors, and entities within sectors.[903] Presently, there are ten sectors under the Framework: industry, building and construction, energy, transport, tourism and culture, health, education, food, waste, and commercial activities.[904]

On September 10, 2005, Abu Dhabi enacted Law No. 21 "For Waste Management in the Emirate of Abu Dhabi" (Law No. 21). Law No. 21 directed the EAD, as the competent authority, to "assume enhancement of waste management within the Emirate."[905] In the context of Law No. 21, this "enhancement" took the form of reducing the volume of waste generated, recycling and reuse of waste, providing various treatment solutions, and setting out priorities and best practices for disposal of the wastes.[906]

Law No. 21 also set out the following set of eight ambitious goals for waste management in the Emirate:[907]

1. Environmental permitting of facilities and activities relevant to waste and permitting of Environmental Services Providers;
2. Review and approval of the methods, mechanisms, and technologies of handling, storage, treatment, and disposal of waste which are proposed by the concerned parties;

3. Assessment of the existing facilities with the relevant concerned parties, permitting of such facilities, and modifying their situations if necessary;
4. Revision and approval of environmental operation, maintenance, and emergency plans of waste storage, treatment, and disposal facilities;
5. Specifying the general requirements for waste management and handling at the Emirate's [sic] level through coordination with the concerned parties if necessary. The competent authority shall identify the regulations, codes of practice, and guidelines that cover all the necessary and required procedures for waste management and handling;
6. Review and approval of the regulations, codes of practice, and guidelines prepared by the concerned parties, and ensure compliance with all of the regulations and guidelines;
7. Monitoring compliance with relevant laws, by-laws, and regulations through inspection of the public and private concerned facilities and auditing of relevant regulations and procedures set out by the concerned parties; and
8. Follow-up implementation of this law, in coordination with concerned parties through:
   a. Proposal and follow-up [of] the execution of necessary measures to encourage the private sector to execute waste management projects.
   b. Review of the environmental issues relating to waste and finding appropriate solutions.
   c. Educating the public on health and environmental risks of waste and hazardous wastes.
   d. Forming technical and administrative committees as required for waste management at the level of the Emirate.

Law No. 21 goes on, in Chapter IV, Article 5, to set out the responsibilities for waste generators in the Emirate. Among the key aspects is a reduction in generated waste via the regulations, methods, techniques, and alternatives which the Emirate has approved for the classification, sorting, reuse, or recycling of waste. Furthermore, waste-generating facilities must enforce OSH requirements at all of their locations. While there are no specific regulations for these requirements cited, the language of Article 5 does specify "all relevant regulations, guidelines and codes of practice."[908] The Article goes on to mandate the proper licensing of waste transport vehicles and the "regulation and monitoring" of transport operations, ensuring contracts for waste treatment, transport, storage, and disposal are only written with Environmental Service Providers permitted by the EAD, and ensuring that parties interested in contracting hazardous wastes are furnished with all available data with respect to the description and specifications of the waste.[909]

Chapter V of Law No. 21 addresses the "Responsibilities of the Storage, Treatment and Disposal Facilities" used for waste. The related Article, Article 6, is surprisingly brief in its content, only providing general guidelines for "concerned and private parties desirous of providing [such] facilities."[910] These include obtaining preliminary approval from the competent authority, preparing a design report for a new, modified, or closed facility – demonstrating how the facility complies with related laws – and obtaining approval/providing an environmental impact assessment, while also obtaining an environmental permit, for new installations and facilities.[911]

Chapter VI of Law No. 21 sets forth the Emirate's requirements for the "Responsibilities of Environmental Service Providers," which are perhaps necessarily more developed than those in Chapter V. Among the key provisions of this Chapter are:[912]

- Obtain a permit from the competent authority before beginning the activity;
- Provide the "material capabilities and human resources" to appropriately manage permitted wastes;
- Work to reduce generated waste and to sort, reuse, and recycle wastes;
- Prepare and apply plans for meeting OSH requirements and for managing pollution, spills, accidents, and related environmental situations;
- Conduct training on waste handling, particularly with respect to managing emergencies; and
- Retain records showing the types of wastes handled, their sources and quantities, any applied treatment(s), pre- and post-treatment analyses, and "waste recipient for disposal" [*Author's note:* it is likely that, from the context of the Chapter, the word should be "receipt."]

10

Concluding Thoughts

The Middle East is an area of the globe which, historically, has not been well understood with respect to chemical substance regulations. While part of this may be due to a relative lack of regulations published in English, it may also be attributed to the facts that the region is economically diverse, and it includes countries at various stages of economic development with vastly different endowments of natural resources.[913] Additionally, the Middle Eastern region's influence in the global economic system historically has been weak.

> Political fragmentation, recurring conflicts, and authoritarian rule have hampered the development of democratic institutions and remain major obstacles to economic reform... The region performs poorly in the areas of civil and political freedoms, gender equality, and, more generally, opportunities for the full development of human capabilities and knowledge.[914]

Full and equal participation in areas of economics and politics, as well as unfettered access to information, are major precursors to the development of a strong national regulatory structure, which would include the management of chemical substances. The sharing of ideas across cultures and borders provides citizens with not only information and context to propose regulatory changes, but also the means by which they can actively communicate these ideas in their own political environments.

> In most countries, elections... are becoming more open and meaningful, and the political leadership is becoming more aware of the need for political reform... [this] reflects the impact of the citizens' vastly expanded access to diverse sources of information as well as internal and external pressures.[915]

Without such stability, any substantive attempt at such will necessarily fall short.

We have observed instances in which cities and regions of various countries have taken a much more direct and stringent role in chemical substance management. The city of Abu Dhabi and the province of Sindh, among others, have established a wide range of in-depth regulations which apply to operations within their area of jurisdiction. These

Chemical Regulation in the Middle East, First Edition. Michael S. Wenk.
© 2018 John Wiley & Sons Ltd. Published 2018 by John Wiley & Sons Ltd.

actions, as well as national movements to incorporate global standards and guidelines, such as the UN's Dangerous Goods regulation and RoHS, speak to an ongoing interest in establishing strong regulatory schemes.

While there are certainly areas for further development in the Middle East with respect to chemical regulations, we have seen a tremendous amount of legislation already implemented, with much more on the horizon. As such, the area appears well positioned for even further progress in the coming years.

Endnotes

1 Szeps-Znaider, T. (2009, November 5). Chemical Control Legislation in the Middle East: Varied and Evolving. *Environmental Business Journal*, XXII(5), pp. 9–12.

2 Ibid.

3 Piccolo, C. (n.d.). The Hejira, or Islamic Calendar. Retrieved from https://www .hziegler.com/articles/heijira-islamic-calendar.html.

4 Ibid.

5 Prime Minister's Decree No. 338 of the Year 1995, Promulgating the Executive Regulations of the Law for the Environment, Law No. 4 for 1994. Retrieved from http://faolex.fao.org/docs/texts/egy4986E.doc.

6 Foundations and Objectives (n.d.). Retrieved from http://www.gcc-sg.org/eng/ index895b.html?action=Sec-Show&ID=3.

7 Gulf Cooperation Council (GCC) and the EU (02/05/2016). Retrieved from http://eeas .europa.eu/gulf_cooperation/index_en.htm.

8 Lee, J. and Attar, Z. (December 2013/January 2014). Paints Regulation in the US and GCC. Chemical Watch Global Business Briefing. Retrieved from https:// chemicalwatch.com/17746/paints-regulation-in-the-us-and-gcc.

9 Ibid.

10 EU Relations with the Gulf Cooperation Council (GCC) (n.d.). Retrieved from http://eeas.europa.eu/gulf_cooperation/index_en.htm.

11 Foundations and Objectives (n.d.). Retrieved from http://www.gcc-sg.org/eng/ index895b.html?action=Sec-Show&ID=3.

12 Ibid.

13 Bergenas, J. (2010, December). A Piece of the Global Puzzle. The Role of the Gulf Cooperation Council and the League of Arab States in Implementing Resolution 1540. Stimson Center, Washington, D.C., p. 20.

14 GSO Technical Regulation on Toys: Second Edition – BD-131704-01. Retrieved from https://www.gso.org.sa/gso-website/gso-website/activities/conformity/technical-regulations-and-guides/bd-131704-01-en-08-09-13.pdf.

15 GSO Technical Regulation on Toys: Second Edition – BD-131704-01. Retrieved from https://www.gso.org.sa/gso-website/gso-website/activities/conformity/technical-regulations-and-guides.

16 General Regulations of Environment in the GCC States (1997). Retrieved from https:// www3.nd.edu/~ggoertz/rei/rei880/rei880.147tt1.pdf.

17 Ibid.

18 Ibid.

19 Ibid.

20 Ibid.

21 Ibid.

22 Ibid.

23 Ibid.

24 Ibid.

25 Ibid.

26 Ibid.

27 Ibid.

28 Ibid.

29 Ibid.

30 Ibid.

31 Ibid.

32 Szeps-Znaider, T. (2009, November 5). Chemical Control Legislation in the Middle East: Varied and Evolving. *Environmental Business Journal*, XXII(5), pp. 9–12.

33 CEFIC (May 2011). Country Sheet: Global Emerging Regulations Issue Team – Middle East. Retrieved from www.coatings.org.uk/Media/Download.aspx?MediaId=2366.

34 Ibid.

35 Lee, J. and Attar, Z. (December 2013/January 2014). Paints Regulation in the US and GCC. Chemical Watch Global Business Briefing. Retrieved from https://chemicalwatch.com/17746/paints-regulation-in-the-us-and-gcc.

36 Common System for the Management of Hazardous Chemicals in the Gulf Cooperation Council for the Arab States of the Gulf (2002).

37 Ibid.

38 Ibid.

39 Ibid.

40 Division 4.1 Flammable Solids. United Nations Recommendations on the Transport of Dangerous Goods Model Regulations. 19th Revised Edition. 2015.

41 Common System for the Management of Hazardous Chemicals in the Gulf Cooperation Council for the Arab States of the Gulf (2002).

42 Ibid.

43 Ibid.

44 Ibid.

45 Ibid.

46 Ibid.

47 Ibid.

48 Ibid.

49 CEFIC (May 2011). Country Sheet: Global Emerging Regulations Issue Team – Middle East. Retrieved from www.coatings.org.uk/Media/Download.aspx?MediaId=2366.

50 Ibid.

51 Common System for the Management of Hazardous Chemicals in the Gulf Cooperation Council for the Arab States of the Gulf (2002).

52 Ibid.

53 Ibid.

54 Ibid.

55 Ibid.

56 Ibid.

57 CEFIC (May 2011). Country Sheet: Global Emerging Regulations Issue Team – Middle East. Retrieved from www.coatings.org.uk/Media/Download.aspx?MediaId=2366.

58 Lee, J. and Attar, Z. (December 2013/January 2014). Paints Regulation in the US and GCC. Chemical Watch Global Business Briefing. Retrieved from https://chemicalwatch.com/17746/paints-regulation-in-the-us-and-gcc.

59 Attar, Z. (2013, September 26). Hazard Communication for the Middle East and Africa. Retrieved from https://schc.memberclicks.net/assets/meetings/fall2013/zeina_attar.ppt.pdf.

60 Ibid.

61 Ibid.

62 Ibid.

63 Pesticides Act of Cooperation Council for the Arab States of the Gulf. Retrieved from http://extwprlegs1.fao.org/docs/pdf/gcc87544.pdf.

64 Law No. 21 of 2009 Approving the Pesticides Act in the Countries of the Cooperation Council for the Arab States of the Gulf. http://extwprlegs1.fao.org/docs/pdf/kuw100469.pdf.

65 Pesticides Act of Cooperation Council for the Arab States of the Gulf. https://www.ecolex.org/details/legislation/pesticides-act-of-cooperation-council-for-the-arab-states-of-the-gulf-lex-faoc087544/.

66 Pesticides Act of Cooperation Council for the Arab States of the Gulf. http://extwprlegs1.fao.org/docs/pdf/gcc87544eng.pdf.

67 Ibid.

68 Anonymous (2016, July 28). Advancing HSE Standards in the GCC. Retrieved from https://antarisconsulting.wordpress.com/2016/07/28/advancing-hse-standards-in-the-gcc/.

69 Ibid.

70 GSO 209 (1994). Industrial Safety and Health Regulations – Part 3: Occupational Health and Environmental Control. Retrieved from https://ia801904.us.archive.org/19/items/gso.209.e.1994/gso.209.e.1994.pdf.

71 GSO 214 (1994). Industrial Safety and Health Regulations Equipment – Materials Handling. Retrieved from https://ia601900.us.archive.org/1/items/gso.214.e.1994/gso.214.e.1994.pdf.

72 Ibid.

73 Kingdom of Bahrain. The Supreme Council for Environment (2012). Kingdom of Bahrain National Profile to Assess the National Infrastructure for Chemical Safety. Retrieved from http://www2.unitar.org/cwm/publications/cw/np/np_pdf/Kingdom of Bahrain_National_Profile_update.pdf.

74 Ibid.

75 Ibid.

76 Ibid.

77 Ibid.

78 Bahrain (2017). CIA World Factbook. Retrieved from https://www.cia.gov/library/publications/the-world-factbook/geos/ba.html.

79 Ibid.

80 Ibid.

81 Ibid.

82 Ibid.

83 Kingdom of Bahrain. The Supreme Council for Environment (2012). Kingdom of Bahrain National Profile to Assess the National Infrastructure for Chemical Safety. Retrieved from http://www2.unitar.org/cwm/publications/cw/np/np_pdf/Kingdom of Bahrain_National_Profile_update.pdf.

84 Ibid.

85 Ibid.

86 Ibid.

87 Ibid.

88 Ibid.

89 Ibid.

90 Ibid.

91 Ibid.

92 CEFIC (May 2011). Country Sheet: Global Emerging Regulations Issue Team – Middle East. Retrieved from www.coatings.org.uk/Media/Download.aspx?MediaId=2366.

93 Kingdom of Bahrain. The Supreme Council for Environment (2012). Kingdom of Bahrain National Profile to Assess the National Infrastructure for Chemical Safety. Retrieved from http://www2.unitar.org/cwm/publications/cw/np/np_pdf/Kingdom of Bahrain_National_Profile_update.pdf.

94 Ahmad, R. (2015, October 25). Environmental Legislations in Kingdom of Bahrain. Retrieved from http://www.ecomena.org/environmental-legislations-Kingdom of Bahrain/.

95 Ibid.

96 Ibid.

97 Legislative Decree No. 21 of 1996 in Respect with the Environment.

98 Ibid.

99 Ibid.

100 Ibid.

101 Ibid.

102 Ibid.

103 Ibid.

104 Ibid.

105 Ibid.

106 Ibid.

107 Ibid.

108 Ibid.

109 Ibid.

110 Resolution No. 1 of 1999 on the Control of Substances that Deplete the Ozone Layer. Retrieved from https://www.ecolex.org/details/legislation/resolution-no-1-of-1999-on-the-control-of-substances-that-deplete-the-ozone-layer-lex-faoc069777/.

111 Kingdom of Bahrain. The Supreme Council for Environment (2012). Kingdom of Bahrain National Profile to Assess the National Infrastructure for Chemical Safety. Retrieved from http://www2.unitar.org/cwm/publications/cw/np/np_pdf/Kingdom of Bahrain_National_Profile_update.pdf.

112 Ibid.

113 Ibid.

114 Ibid.

115 Ibid.

116 Resolution No. 4 of 2006 (May 31). On the Management of Hazardous Chemicals. In Official Gazette of Kingdom of Bahrain, Issue 2741.

117 Attar, Z. (2013, September 26). Hazard Communication for the Middle East and Africa. Retrieved from https://schc.memberclicks.net/assets/meetings/fall2013/ zeina_attar.ppt.pdf.

118 Ibid.

119 Kingdom of Bahrain. The Supreme Council for Environment (2012). Kingdom of Bahrain National Profile to Assess the National Infrastructure for Chemical Safety. Retrieved from http://www2.unitar.org/cwm/publications/cw/np/np_pdf/Kingdom of Bahrain_National_Profile_update.pdf.

120 Ibid.

121 Legislative Decree No. 11 of 1989 on Pesticides. Retrieved from https://www.ecolex .org/details/legislation/legislative-decree-no-11-of-1989-on-pesticides-lex-faoc072227/.

122 Kingdom of Bahrain. The Supreme Council for Environment (2012). Kingdom of Bahrain National Profile to Assess the National Infrastructure for Chemical Safety. Retrieved from http://www2.unitar.org/cwm/publications/cw/np/np_pdf/Kingdom of Bahrain_National_Profile_update.pdf.

123 Ibid.

124 Ibid.

125 Ibid.

126 Ibid.

127 Pesticides Registration. Retrieved from http://websrv.municipality.gov.bh/agri/pages/ pesticide_registeration_en.jsp.

128 Ibid.

129 Kingdom of Bahrain. The Supreme Council for Environment (2012). Kingdom of Bahrain National Profile to Assess the National Infrastructure for Chemical Safety. Retrieved from http://www2.unitar.org/cwm/publications/cw/np/np_pdf/Kingdom of Bahrain_National_Profile_update.pdf.

130 Matooq, A. and Suliman, S. (2013). Performance Measurement of Occupational Safety and Health: Model for Bahrain Inspectorates. *Journal of Safety Engineering*, 2(2), pp. 26–38.

131 Kingdom of Bahrain. The Supreme Council for Environment (2012). Kingdom of Bahrain National Profile to Assess the National Infrastructure for Chemical Safety. Retrieved from http://www2.unitar.org/cwm/publications/cw/np/np_pdf/Kingdom of Bahrain_National_Profile_update.pdf.

132 Ibid.

133 Matooq, A. and Suliman, S. (2013). Performance Measurement of Occupational Safety and Health: Model for Bahrain Inspectorates. *Journal of Safety Engineering*, 2(2), pp. 26–38.

134 Ibid.

135 Ibid.

136 Ibid.

137 Ibid.

138 Kingdom of Bahrain. The Supreme Council for Environment (2012). Kingdom of Bahrain National Profile to Assess the National Infrastructure for Chemical Safety.

Retrieved from http://www2.unitar.org/cwm/publications/cw/np/np_pdf/Kingdom of Bahrain_National_Profile_update.pdf.

139 Ibid.

140 Ibid.

141 Ibid.

142 Ibid.

143 Resolution (3) (2006). Hazardous Waste Management. Retrieved from http://archive .basel.int/legalmatters/natleg/byparties/dnn-frmbody.php?partyId=11.

144 Ibid.

145 Ibid.

146 Ibid.

147 Ibid.

148 Ibid.

149 Ibid.

150 Ibid.

151 HRH Premier issues [sic] Edict on Safety and Occupational Health Council (2015, January 7). Retrieved from http://bna.bh/portal/en/news/648668.

152 Prime Ministerial Edict No. (2) of 2015 With Respect to the Occupational Safety and Health Council. (2015, January 8). The Official Gazette, Issue No. 3191. Retrieved from http://www.stc-bahrain.com/images/OccSafetyandHealth.pdf.

153 Ibid.

154 Ibid.

155 Ibid.

156 Draft Ministerial Order No. 7 for the Year 2002 in Respect of Hazardous Waste Management. Retrieved from http://extwprlegs1.fao.org/docs/texts/bah68177E.doc.

157 Ibid.

158 Ibid.

159 Ibid.

160 Ibid.

161 Attar, Z. (2013, September 26). Hazard Communication for the Middle East and Africa. Retrieved from https://schc.memberclicks.net/assets/meetings/fall2013/ zeina_attar.ppt.pdf.

162 Kingdom of Bahrain. The Supreme Council for Environment (2012). Kingdom of Bahrain National Profile to Assess the National Infrastructure for Chemical Safety. Retrieved from http://www2.unitar.org/cwm/publications/cw/np/np_pdf/Kingdom of Bahrain_National_Profile_update.pdf.

163 Attar, Z. (2013, September 26). Hazard Communication for the Middle East and Africa. Retrieved from https://schc.memberclicks.net/assets/meetings/fall2013/ zeina_attar.ppt.pdf.

164 Ibid.

165 El Zarka, M. (1999, January). National Profile for the Management of Chemicals in Egypt. Egyptian Environmental Affairs Agency. Retrieved from http://cwm.unitar.org/ national-profiles/publications/cw/np/np_pdf/Egypt_National_Profile.pdf.

166 Egypt (2017). CIA World Factbook. Retrieved from https://www.cia.gov/library/ publications/the-world-factbook/geos/eg.html.

167 Ibid.

168 Ibid.

169 Ibid.

170 Ibid.

171 Ibid.

172 Ibid.

173 Ibid.

174 Ibid.

175 Ibid.

176 Anonymous (2006). Egypt. Guide to Foreign and International Legal Citations. *Journal of International Law and Politics.* New York University School of Law.

177 Ibid.

178 Ibid.

179 Hanafi Islam (n.d.). Retrieved from https://www.globalsecurity.org/military/intro/ islam-hanafi.htm.

180 Egypt (2017). CIA World Factbook. Retrieved from https://www.cia.gov/library/ publications/the-world-factbook/geos/eg.html.

181 Ibid.

182 Ibid.

183 El Zarka, M. (1999, January). National Profile for the Management of Chemicals in Egypt. Egyptian Environmental Affairs Agency. Retrieved from http://cwm.unitar.org/ national-profiles/publications/cw/np/np_pdf/Egypt_National_Profile.pdf.

184 Ibid.

185 Official Journal, Issue No. 51, February 28, 1995.

186 Prime Minister's Decree No. 338 of the Year 1995, Promulgating the Executive Regulations of the Law for the Environment, Law No. 4 for 1994. Retrieved from http://faolex.fao.org/docs/texts/egy4986E.doc.

187 Ministry of Environment, Egyptian Environmental Affairs Agency (n.d.). About MSEA – EEAA. Retrieved from http://www.eeaa.gov.eg/english/main/about_detail .asp.

188 El Zarka, M. (1999, January). National Profile for the Management of Chemicals in Egypt. Egyptian Environmental Affairs Agency. Retrieved from http://cwm.unitar.org/ national-profiles/publications/cw/np/np_pdf/Egypt_National_Profile.pdf.

189 Ministry of Environment, Egyptian Environmental Affairs Agency (n.d.). About MSEA – EEAA. Retrieved from http://www.eeaa.gov.eg/english/main/about_detail .asp.

190 Prime Minister's Decree No. 338 of the Year 1995, Promulgating the Executive Regulations of the Law for the Environment, Law No. 4 for 1994. Retrieved from http://faolex.fao.org/docs/texts/egy4986E.doc.

191 Ministry of Environment, Egyptian Environmental Affairs Agency (n.d.). Environmental Management – Chemicals Management Projects. Retrieved from http://www.eeaa.gov.eg/english/main/about_detail.asp.

192 Anonymous (n.d.). Regional Workshop on Chemical Hazard Communication and GHS Implementation for Arab Countries. Retrieved from http://cwm.unitar.org/ publications/publications/event/ghs_arab_workshop_2006/Arab_Regional_GHS_ Workshop_Report_final_rev.pdf.

193 CEFIC (May 2011). Country Sheet: Global Emerging Regulations Issue Team – Middle East. Retrieved from www.coatings.org.uk/Media/Download.aspx?MediaId=2366.

194 El Zarka, M. (1999, January). National Profile for the Management of Chemicals in Egypt. Egyptian Environmental Affairs Agency. Retrieved from http://cwm.unitar.org/national-profiles/publications/cw/np/np_pdf/Egypt_National_Profile.pdf.

195 Ibid.

196 Ibid.

197 Japan International Cooperation Agency (JICA) (2002, February). Country Profile on Environment Egypt. Retrieved from https://www.jica.go.jp/english/.

198 CEFIC (May 2011). Country Sheet: Global Emerging Regulations Issue Team – Middle East. Retrieved from www.coatings.org.uk/Media/Download.aspx?MediaId=2366.

199 Anonymous (n.d.). Regional Workshop on Chemical Hazard Communication and GHS Implementation for Arab Countries. Retrieved from http://cwm.unitar.org/publications/publications/event/ghs_arab_workshop_2006/Arab_Regional_GHS_Workshop_Report_final_rev.pdf.

200 Egyptian Environmental Affairs Agency (EEAA). Chapter 4: Legal Instruments and Non-Regulatory Mechanisms for Managing Chemicals. Retrieved from www.eeaa.gov.eg/ehsims/main/NCP/Chapter%204.doc.

201 Soliman, A. (2011, November 2). Industrial Licensing. Egyptian Regulatory Reform & Development Activity. Retrieved from http://www.errada.gov.eg/index_en.php?op=show_feature_details_en&id=8&start=12&type=2.

202 Egyptian Environmental Affairs Agency (EEAA). Chapter 4: Legal Instruments and Non-Regulatory Mechanisms for Managing Chemicals. Retrieved from www.eeaa.gov.eg/ehsims/main/NCP/Chapter%204.doc.

203 Ibid.

204 Anonymous (n.d.). Chapter 4: Legal Instruments and Non-Regulatory Mechanisms for Managing Chemicals. Egyptian Hazardous Substance Information and Management System. Retrieved from www.eeaa.gov.eg/ehsims/main/NCP/Chapter%204.doc.

205 Ibid.

206 Ibid.

207 Ibid.

208 Ibid.

209 Ibid.

210 Ibid.

211 Japan International Cooperation Agency (JICA) (2002, February). Country Profile on Environment Egypt. Retrieved from https://www.jica.go.jp/english/.

212 Prime Minister's Decree No. 338 of the Year 1995, Promulgating the Executive Regulations of the Law for the Environment, Law No. 4 for 1994. Retrieved from http://faolex.fao.org/docs/texts/egy4986E.doc.

213 Ibid.

214 Ibid.

215 Ibid.

216 Ibid.

217 Ibid.

218 Ibid.

219 Ibid.

220 Ibid.

221 Anonymous (n.d.). Chapter 4: Legal Instruments and Non-Regulatory Mechanisms for Managing Chemicals. Egyptian Hazardous Substance Information and Management System. Retrieved from www.eeaa.gov.eg/ehsims/main/NCP/Chapter%204.doc.

222 Ibid.

223 Prime Minister's Decree No. 338 of the Year 1995, Promulgating the Executive Regulations of the Law for the Environment, Law No. 4 for 1994. Retrieved from http://faolex.fao.org/docs/texts/egy4986E.doc.

224 El Zarka, M. (1999, January). National Profile for the Management of Chemicals in Egypt. Egyptian Environmental Affairs Agency. Retrieved from http://cwm.unitar.org/national-profiles/publications/cw/np/np_pdf/Egypt_National_Profile.pdf.

225 Ibid.

226 Ibid.

227 Ibid.

228 Ibid.

229 Egypt. Ministry of Agricultural and Land Reclamation. Agricultural Pesticides Committee. Retrieved from http://www.apc.gov.eg/EN/.

230 Ministerial Decree No. 1018 of the Year 2013, Concerning Registration, Handling and Use of Agricultural Pesticides in Egypt. Retrieved from www.apc.gov.eg/Files/M.D .1018-2013.pdf.

231 Ibid.

232 Ibid.

233 Ibid.

234 Ibid.

235 Ibid.

236 Ibid.

237 Ibid.

238 Ibid.

239 Ibid.

240 About Us. Central Agricultural Pesticide Laboratory (CAPL). Retrieved from http://www.arc.sci.eg/InstsLabs/Default.aspx?OrgID=17&lang=en.

241 Ministerial Decree No. 1018 of the Year 2013, Concerning Registration, Handling and Use of Agricultural Pesticides in Egypt. Retrieved from www.apc.gov.eg/Files/M.D .1018-2013.pdf.

242 Ibid.

243 Ibid.

244 Ibid.

245 Ibid.

246 Ibid.

247 Ibid.

248 Ibid.

249 Ibid.

250 Prime Minister's Decree No. 338 of the Year 1995, Promulgating the Executive Regulations of the Law for the Environment, Law No. 4 for 1994. Retrieved from http://faolex.fao.org/docs/texts/egy4986E.doc.

251 Abo El Ata, G.A. and Nahmias, M. (2005, January). Occupational Safety and Health in Egypt – A National Profile. International Labor Organization. Retrieved from http://www.ilo.org/safework/countries/africa/egypt/lang--en/index.htm.

252 Ibid.

253 Ibid.

254 Ibid.

255 Ibid.

256 Ibid.

257 Ibid.

258 Ibid.

259 Ibid.

260 Ibid.

261 Ibid.

262 Ibid.

263 Ibid.

264 Ibid.

265 Ibid.

266 Ibid.

267 Ibid.

268 Ibid.

269 Ibid.

270 Ibid.

271 Ibid.

272 Anonymous (n.d.). Chapter 4: Legal Instruments and Non-Regulatory Mechanisms for Managing Chemicals. Egyptian Hazardous Substance Information and Management System. Retrieved from www.eeaa.gov.eg/ehsims/main/NCP/Chapter%204.doc.

273 Ibid.

274 Egyptian Environmental Affairs Agency (EEAA). Chapter 4: Legal Instruments and Non-Regulatory Mechanisms for Managing Chemicals. Retrieved from www.eeaa.gov .eg/ehsims/main/NCP/Chapter%204.doc.

275 Prime Minister's Decree No. 338 of the Year 1995, Promulgating the Executive Regulations of the Law for the Environment, Law No. 4 for 1994. Retrieved from http://faolex.fao.org/docs/texts/egy4986E.doc.

276 Ramadan, A.R. and Nadim, A.H. (September, 29–October, 1, 2004). Hazardous Waste Management in Egypt: Status and Challenges. Proceedings of the 2nd International Conference on Waste Management and the Environment, V. Popov, H. Itoh, C.A. Brebbia & S. Kungolos (Eds). WIT Press, 2004.

277 Ibid.

278 Ibid.

279 Ibid.

280 Ibid.

281 Ibid.

282 Ibid.

283 Ibid.

284 Ibid.

285 Ibid.

286 CEFIC (May 2011). Country Sheet: Global Emerging Regulations Issue Team – Middle East. Retrieved from www.coatings.org.uk/Media/Download.aspx?MediaId=2366.

287 Ibid.

288 CEFIC (May 2012). Global GHS Implementation Task Force – Overview of GHS implementations. Retrieved from http://www.petrochemistry.eu/index.php?option=com_ccnewsletter&task=download&id=504.

289 Anonymous (n.d.). Chapter 4: Legal Instruments and Non-Regulatory Mechanisms for Managing Chemicals. Egyptian Hazardous Substance Information and Management System. Retrieved from www.eeaa.gov.eg/ehsims/main/NCP/Chapter%204.doc.

290 Ibid.

291 Ibid.

292 Ibid.

293 Anonymous (n.d.). Israel country profile. BBC News. Retrieved from http://www.bbc.com/news/world-middle-east-14628835.

294 Creation of Israel, 1948 (n.d.). Retrieved from https://history.state.gov/milestones/1945-1952/creation-israel.

295 Ibid.

296 Israel (2017). CIA World Factbook. Retrieved from https://www.cia.gov/library/publications/the-world-factbook/geos/is.html.

297 Kaplan, E. and Friedman, C. (2009, February 11). Israel's Political System. Council on Foreign Relations. Retrieved from https://www.cfr.org/backgrounder/israels-political-system.

298 Anonymous (2006). Israel. Guide to Foreign and International Legal Citations. *Journal of International Law and Politics*. New York University School of Law.

299 Israel (2017). CIA World Factbook. Retrieved from https://www.cia.gov/library/publications/the-world-factbook/geos/is.html.

300 Ibid.

301 Anonymous (2006). Israel. Guide to Foreign and International Legal Citations. *Journal of International Law and Politics*. New York University School of Law.

302 Ibid.

303 Kaplan, E. and Friedman, C. (2009, February 11). Israel's Political System. Council on Foreign Relations. Retrieved from https://www.cfr.org/backgrounder/israels-political-system.

304 Anonymous (2006). Israel. Guide to Foreign and International Legal Citations. *Journal of International Law and Politics*. New York University School of Law.

305 Anonymous (n.d.). Common Law. Merriam-Webster Dictionary. Retrieved from https://www.merriam-webster.com/dictionary/common%20law.

306 Anonymous (2006). Israel. Guide to Foreign and International Legal Citations. *Journal of International Law and Politics*. New York University School of Law.

307 Israel (2017). CIA World Factbook. Retrieved from https://www.cia.gov/library/publications/the-world-factbook/geos/is.html.

308 Ibid.

309 Ibid.

310 Anonymous (2006). Israel. Guide to Foreign and International Legal Citations. *Journal of International Law and Politics*. New York University School of Law.

311 Szeps-Znaider, T. (2009, May). Chemicals Regulation in the Middle East – An Overview. Retrieved from https://chemicalwatch.com/2221/chemicals-regulation-in-the-middle-east-an-overview.

312 Szeps-Znaider, T. (2009, November 5). Chemical Control Legislation in the Middle East: Varied and Evolving. *Environmental Business Journal*, XXII(5), pp. 9–12.

313 Golstein-Galperin, R. (2010, October 7). Regulation in an Era of Global Governance – Israel's Chemicals Regulatory Regime Following the Accession to the OECD. Hebrew University of Jerusalem, Faculty of Social Science, Federmann School of Public Policy and Government. Retrieved from http://public-policy.huji.ac.il/ .upload/thesis_EN/Rita_Golstein_Galperin_full.pdf.

314 Ministry of Environmental Protection (2009, April). Chemicals Management in Israel. Retrieved from http://www.sviva.gov.il/English/env_topics/InternationalCooperation/ OECD/Documents/BriefNoteToOECD-ChemicalManagementInIsrael-April2009.pdf.

315 Golstein-Galperin, R. (2010, October 7). Regulation in an Era of Global Governance – Israel's Chemicals Regulatory Regime Following the Accession to the OECD. Hebrew University of Jerusalem, Faculty of Social Science, Federmann School of Public Policy and Government. Retrieved from http://public-policy.huji.ac.il/ .upload/thesis_EN/Rita_Golstein_Galperin_full.pdf.

316 Ibid.

317 Israel National Report: Eighteenth Session of the Commission on Sustainable Development (2010, April). Retrieved from http://www.un.org/esa/dsd/dsd_aofw_ni/ ni_pdfs/NationalReports/israel/Full_text.pdf.

318 Ministry of Environmental Protection (2009, April). Chemicals Management in Israel. Retrieved from http://www.sviva.gov.il/English/env_topics/InternationalCooperation/ OECD/Documents/BriefNoteToOECD-ChemicalManagementInIsrael-April2009.pdf.

319 Golstein-Galperin, R. (2010, October 7). Regulation in an Era of Global Governance – Israel's Chemicals Regulatory Regime Following the Accession to the OECD. Hebrew University of Jerusalem, Faculty of Social Science, Federmann School of Public Policy and Government. Retrieved from http://public-policy.huji.ac.il/ .upload/thesis_EN/Rita_Golstein_Galperin_full.pdf.

320 Israel National Report: Eighteenth Session of the Commission on Sustainable Development (2010, April). Retrieved from http://www.un.org/esa/dsd/dsd_aofw_ni/ ni_pdfs/NationalReports/israel/Full_text.pdf.

321 Ibid.

322 Golstein-Galperin, R. (2010, October 7). Regulation in an Era of Global Governance – Israel's Chemicals Regulatory Regime Following the Accession to the OECD. Hebrew University of Jerusalem, Faculty of Social Science, Federmann School of Public Policy and Government. Retrieved from http://public-policy.huji.ac.il/ .upload/thesis_EN/Rita_Golstein_Galperin_full.pdf.

323 Ibid.

324 Israel National Report: Eighteenth Session of the Commission on Sustainable Development (2010, April). Retrieved from http://www.un.org/esa/dsd/dsd_aofw_ni/ ni_pdfs/NationalReports/israel/Full_text.pdf.

325 Health and Safety (n.d.). Retrieved from http://www.sviva.gov.il/English/Legislation/ Pages/HealthAndSafety.aspx.

326 Golstein-Galperin, R. (2010, October 7). Regulation in an Era of Global Governance – Israel's Chemicals Regulatory Regime Following the Accession to the OECD. Hebrew University of Jerusalem, Faculty of Social Science, Federmann School of Public Policy and Government. Retrieved from http://public-policy.huji.ac.il/ .upload/thesis_EN/Rita_Golstein_Galperin_full.pdf.

327 Israel National Report: Eighteenth Session of the Commission on Sustainable Development (2010, April). Retrieved from http://www.un.org/esa/dsd/dsd_aofw_ni/ni_pdfs/NationalReports/israel/Full_text.pdf.

328 Hazardous Substances Law (1993). Retrieved from http://www.sviva.gov.il/English/Legislation/Documents/Hazardous%20Substances%20Laws%20and%20Regulations/HazardousSubstancesLaw1993.pdf.

329 Szeps-Znaider, T. (2009, November 5). Chemical Control Legislation in the Middle East: Varied and Evolving. *Environmental Business Journal*, XXII(5), pp. 9–12.

330 Hazardous Substances Law (1993). Retrieved from http://www.sviva.gov.il/English/Legislation/Documents/Hazardous%20Substances%20Laws%20and%20Regulations/HazardousSubstancesLaw1993.pdf.

331 Ibid.

332 Ibid.

333 Ibid.

334 Ibid.

335 Ibid.

336 Licensing of Businesses Law (5278-1968). Retrieved from http://financeisrael.mof.gov.il/FinanceIsrael/Docs/En/legislation/Others/5728_1968_Licensing%20of_Businesses_Law.pdf.

337 Ibid.

338 Ibid.

339 Israel National Report: Eighteenth Session of the Commission on Sustainable Development (2010, April). Retrieved from http://www.un.org/esa/dsd/dsd_aofw_ni/ni_pdfs/NationalReports/israel/Full_text.pdf.

340 Ibid.

341 Ibid.

342 Golstein-Galperin, R. (2010, October 7). Regulation in an Era of Global Governance – Israel's Chemicals Regulatory Regime Following the Accession to the OECD. Hebrew University of Jerusalem, Faculty of Social Science, Federmann School of Public Policy and Government. Retrieved from http://public-policy.huji.ac.il/.upload/thesis_EN/Rita_Golstein_Galperin_full.pdf.

343 Ibid.

344 Ibid.

345 Anonymous (n.d.). Health and Safety. Israel Ministry of Environmental Protection. Retrieved from http://www.sviva.gov.il/English/Legislation/Pages/HealthAndSafety.aspx.

346 Ibid.

347 Licensing of Business Regulations (Hazardous Plants) 5753-1993. Retrieved from http://www.sviva.gov.il/English/Legislation/Documents/Licensing%20of%20Businesses%20Laws%20and%20Regulations/LicensingOfBusinessesRegulations-HazardousPlants-1993.pdf.

348 Ibid.

349 Ibid.

350 Anonymous (n.d.). Health and Safety. Israel Ministry of Environmental Protection. Retrieved from http://www.sviva.gov.il/English/Legislation/Pages/HealthAndSafety.aspx.

351 Licensing of Business Regulations (Hazardous Plants) 5753-1993. Retrieved from
http://www.sviva.gov.il/English/Legislation/Documents/Licensing%20of
%20Businesses%20Laws%20and%20Regulations/LicensingOfBusinessesRegulations-
HazardousPlants-1993.pdf.

352 Anonymous (n.d.). Health and Safety. Israel Ministry of Environmental Protection.
Retrieved from http://www.sviva.gov.il/English/Legislation/Pages/HealthAndSafety
.aspx.

353 Licensing of Business Regulations (Hazardous Plants) 5753-1993. Retrieved from
http://www.sviva.gov.il/English/Legislation/Documents/Licensing%20of
%20Businesses%20Laws%20and%20Regulations/LicensingOfBusinessesRegulations-
HazardousPlants-1993.pdf.

354 Ibid.

355 Ibid.

356 Ibid.

357 Hazardous Substances Regulations (Criteria for Determining Validity Period of
Permits), 5763-2003. Retrieved from http://www.sviva.gov.il/English/Legislation/
Documents/Hazardous%20Substances%20Laws%20and%20Regulations/
HazardousSubstancesRegulations-CriteriaForDeterminingPermitValidityPeriod-2003
.pdf.

358 Ibid.

359 Ibid.

360 Ibid.

361 Ibid.

362 Ibid.

363 Ibid.

364 Ibid.

365 Ibid.

366 Ibid.

367 Anonymous (n.d.). Health and Safety. Israel Ministry of Environmental Protection.
Retrieved from http://www.sviva.gov.il/English/Legislation/Pages/HealthAndSafety
.aspx.

368 Ibid.

369 Hazardous Substances Regulation (Classification and Exemption) 5756-1996.
Retrieved from http://www.sviva.gov.il/English/Legislation/Documents/Hazardous
%20Substances%20Laws%20and%20Regulations/HazardousSubstancesRegulations-
ClassificationAndExemption-1996.pdf.

370 Ibid.

371 Ibid.

372 Golstein-Galperin, R. (2010, October 7). Regulation in an Era of Global
Governance – Israel's Chemicals Regulatory Regime Following the Accession to the
OECD. Hebrew University of Jerusalem, Faculty of Social Science, Federmann School
of Public Policy and Government. Retrieved from http://public-policy.huji.ac.il/
.upload/thesis_EN/Rita_Golstein_Galperin_full.pdf.

373 Ibid.

374 Ibid.

375 Israel National Report: Eighteenth Session of the Commission on Sustainable Development (2010, April). Retrieved from http://www.un.org/esa/dsd/dsd_aofw_ni/ni_pdfs/NationalReports/israel/Full_text.pdf.

376 Ibid.

377 Anonymous (n.d.). Health and Safety. Israel Ministry of Environmental Protection. Retrieved from http://www.sviva.gov.il/English/Legislation/Pages/HealthAndSafety.aspx.

378 Ibid.

379 Hazardous Substances Regulations (Registration of Preparations for the Control of Pests Harmful to Humans), 5754-1994. Retrieved from http://www.sviva.gov.il/English/Legislation/Documents/Hazardous%20Substances%20Laws%20and%20Regulations/HazardousSubstancesRegulations-RegistrationOfPreparationsForControlOfHarmfulPests-1994.pdf.

380 Council Regulation (EU) Regulation (EC) No. 1907/2006 of the European Parliament and of the Council of December 18, 2006 concerning the Registration, Evaluation, Authorization and Restriction of Chemicals (REACH), establishing a European Chemicals Agency, amending Directive 1999/45/EC and repealing Council Regulation (EEC) No. 793/93 and Commission Regulation (EC) No. 1488/94 as well as Council Directive 76/769/EEC and Commission Directives 91/155/EEC, 93/67/EEC, 93/105/EC, and 2000/21/EC. *Official Journal of the European Union*, L391, pp. 1–849.

381 Hazardous Substances Regulations (Registration of Preparations for the Control of Pests Harmful to Humans), 5754-1994. Retrieved from http://www.sviva.gov.il/English/Legislation/Documents/Hazardous%20Substances%20Laws%20and%20Regulations/HazardousSubstancesRegulations-RegistrationOfPreparationsForControlOfHarmfulPests-1994.pdf.

382 Ibid.

383 Ibid.

384 Federal Insecticide, Fungicide and Rodenticide Act, 40 C.F.R. §156.10(g)(i)(2)(ii) (2017).

385 Hazardous Substances Regulations (Registration of Preparations for the Control of Pests Harmful to Humans), 5754-1994. Retrieved from http://www.sviva.gov.il/English/Legislation/Documents/Hazardous%20Substances%20Laws%20and%20Regulations/HazardousSubstancesRegulations-RegistrationOfPreparationsForControlOfHarmfulPests-1994.pdf.

386 Ibid.

387 Ibid.

388 Ibid.

389 Anonymous (n.d.). Health and Safety. Israel Ministry of Environmental Protection. Retrieved from http://www.sviva.gov.il/English/Legislation/Pages/HealthAndSafety.aspx.

390 Ibid.

391 Hazardous Substances Regulations (Registration of Preparations for the Control of Pests Harmful to Humans), 5754-1994. Retrieved from http://www.sviva.gov.il/English/Legislation/Documents/Hazardous%20Substances%20Laws%20and%20Regulations/HazardousSubstancesRegulations-RegistrationOfPreparationsForControlOfHarmfulPests-1994.pdf.

392 Ibid.

393 Anonymous (n.d.). Health and Safety. Israel Ministry of Environmental Protection. Retrieved from http://www.sviva.gov.il/English/Legislation/Pages/HealthAndSafety .aspx.

394 Ibid.

395 Safety and Health Administration. Ministry of Economy and Industry. Retrieved from http://www.economy.gov.il/English/About/Units/Pages/SafetyAndHealth.aspx.

396 Ibid.

397 Ibid.

398 Work Safety Ordinance 5730-1970. Retrieved from http://www.ilo.org/dyn/natlex/ docs/ELECTRONIC/36158/97939/F1486776006/ISR36158.pdf.

399 Ibid.

400 Ibid.

401 Ibid.

402 Ibid.

403 General Information. Israel Institute for Occupational Safety and Hygiene. Retrieved from https://www.osh.org.il/eng/eng_site/about_iiosh/.

404 Licensing of Business Regulations (Disposal of Hazardous Substances Waste), 5751-1990. Retrieved from http://www.sviva.gov.il/English/Legislation/Documents/ Licensing%20of%20Businesses%20Laws%20and%20Regulations/ LicensingOfBusinessesRegulations-DisposalOfHazmatWaste-1990.pdf.

405 Ibid.

406 Ibid.

407 Ibid.

408 Hazardous Substances Regulation (Import and Export of Hazardous Substances Waste), 5754-1994. Retrieved from http://www.sviva.gov.il/English/Legislation/ Documents/Hazardous%20Substances%20Laws%20and%20Regulations/ HazardousSubstancesRegulations-ImportAndExportOfHazardousSubstancesWaste- 1994.pdf.

409 Ibid.

410 Ibid.

411 Szeps-Znaider, T. (2009, November 5). Chemical Control Legislation in the Middle East: Varied and Evolving. *Environmental Business Journal*, XXII(5), pp. 9–12.

412 Anonymous (n.d.). Health and Safety. Israel Ministry of Environmental Protection. Retrieved from http://www.sviva.gov.il/English/Legislation/Pages/HealthAndSafety .aspx.

413 Levinson, T. (October 30, 2006). National Legislation – Comparison of Safety Data Sheets for Israel and EU Markets. International Law Office. Retrieved from http:// www.internationallawoffice.com/Newsletters/detail.aspx?g=f8d3e01a-5363-db11- a275-001143e35d55.

414 Work Safety Regulations (Safety Data Sheets, Classification, Packing, Labeling and Marking of Packages), 5758-1998. Retrieved from http://www.sviva.gov.il/English/ Legislation/Documents/Safety%20at%20Work%20Laws%20and%20Regulations/ WorkSafetyRegulations-Packages-1998.pdf.

415 Ibid.

416 Ibid.

417 Ibid.

418 Ibid.

419 Ibid.

420 Work Safety Regulations (Safety Data Sheets, Classification, Packing, Labeling and Marking of Packages), 5758-1998. Retrieved from http://www.sviva.gov.il/English/ Legislation/Documents/Safety%20at%20Work%20Laws%20and%20Regulations/ WorkSafetyRegulations-Packages-1998.pdf.

421 Ibid.

422 Levinson, T. (2006, October 30). National Legislation – Comparison of Safety Data Sheets for Israel and EU Markets. International Law Office. Retrieved from http:// www.internationallawoffice.com/Newsletters/detail.aspx?g=f8d3e01a-5363-db11- a275-001143e35d55.

423 Ibid.

424 Ibid.

425 Work Safety Regulations (Safety Data Sheets, Classification, Packing, Labeling and Marking of Packages), 5758-1998. Retrieved from http://www.sviva.gov.il/English/ Legislation/Documents/Safety%20at%20Work%20Laws%20and%20Regulations/ WorkSafetyRegulations-Packages-1998.pdf.

426 Ibid.

427 Anonymous (n.d.). GHS Implementation – Israel. Retrieved from https://www.unece .org/trans/danger/publi/ghs/implementation_e.html#c53176.

428 Khedr, A. (2010, January/February). Kuwait's Legal System and Legal Research. GlobaLex. New York University. Retrieved from http://www.nyulawglobal.org/ globalex/Kuwait.html.

429 Crystal, J., Sadek, D., Ochsenwald, W. and Anthony, J. (2017, August 15). Kuwait. Retrieved from https://www.britannica.com/place/Kuwait.

430 Kuwait (2017). CIA World Factbook. Retrieved from https://www.cia.gov/library/ publications/the-world-factbook/geos/ku.html.

431 Anonymous (n.d.). The Political System of Kuwait. Helen Ziegler & Associates. Retrieved from https://www.hziegler.com/articles/political-system-of-kuwait.html.

432 Ibid.

433 Ibid.

434 Ibid.

435 Ibid.

436 Anonymous (1999). Law No. 21 of 1999 Establishing the Public Authority for the Environment. *Arab Law Quarterly*, 14(1), pp. 79–87.

437 Anonymous (n.d.). Director Message. Environment Public Authority. Retrieved from https://www.epa.org.kw/article.php?id=21.

438 Anonymous (n.d.). Supreme Council. Environment Public Authority. Retrieved from https://www.epa.org.kw/members.php?list=24.

439 Ibid.

440 Anonymous (1999). Law No. 21 of 1999 Establishing the Public Authority for the Environment. *Arab Law Quarterly*, 14(1), pp. 79–87.

441 Ibid.

442 Anonymous (n.d.). Director Message. Environment Public Authority. Retrieved from https://www.epa.org.kw/article.php?id=21.

443 Anonymous (n.d.). Regional Workshop on Chemical Hazard Communication and GHS Implementation for Arab Countries. Retrieved from http://cwm.unitar.org/

publications/publications/event/ghs_arab_workshop_2006/Arab_Regional_GHS_
Workshop_Report_final_rev.pdf.

444 Anonymous (1995). Law No. 21/1995 Kuwait Environment Public Authority. Further
amended: Law No. 16/1996. The London School of Economics and Political Science:
Grantham Research Institute on Climate Change and the Environment. Retrieved
from http://www.lse.ac.uk/GranthamInstitute/law/law-no-211995-kuwait-
environment-public-authority-further-amended-law-no-161996/.

445 Ibid.

446 Garlaparti, A. and Adivi, B. (n.d.). An Outline of HSE Regulations of State of Kuwait.
American Society of Safety Engineers (ASSE). Retrieved from http://www.assekuwait
.org/download/books/1904151047181585.pdf.

447 Anonymous (1995). Law No. 21/1995 Kuwait Environment Public Authority. Further
amended: Law No. 16/1996. The London School of Economics and Political Science:
Grantham Research Institute on Climate Change and the Environment. Retrieved
from http://www.lse.ac.uk/GranthamInstitute/law/law-no-211995-kuwait-
environment-public-authority-further-amended-law-no-161996/.

448 Ibid.

449 Ibid.

450 Ibid.

451 Anonymous (1999). Law No. 21 of 1999 Establishing the Public Authority for the
Environment. *Arab Law Quarterly*, 14(1), pp. 79–87.

452 Ibid.

453 Ibid.

454 Ibid.

455 Anonymous (1995). Law No. 21/1995 Kuwait Environment Public Authority. Further
amended: Law No. 16/1996. The London School of Economics and Political Science:
Grantham Research Institute on Climate Change and the Environment. Retrieved
from http://www.lse.ac.uk/GranthamInstitute/law/law-no-211995-kuwait-
environment-public-authority-further-amended-law-no-161996/.

456 Environmental Protection Authority (2001, October 2). Decision No. 210/2001
Pertaining to the Executive By-Law of the Law of Environment Public Authority.
Retrieved from http://extwprlegs1.fao.org/docs/pdf/kuw159620E.pdf.

457 Ibid.

458 Ibid.

459 Ibid.

460 Ibid.

461 Ibid.

462 Ibid.

463 Ibid.

464 Ibid.

465 Ibid.

466 Ibid.

467 Anonymous (n.d.). Cosmetic Products. Safety Requirements of Cosmetics and
Personal Care Products. Notification to the World Trade Organization. Retrieved from
https://docs.wto.org/dol2fe/Pages/FE_Search/FE_S_S009-DP.aspx?language=E&
CatalogueIdList=129677&CurrentCatalogueIdIndex=0&FullTextSearch=.

468 Anonymous (2015, January 20). Kuwait Standards and Meteorology Department. Cosmetic Products – Safety Requirements of Cosmetics and Personal Care Products. Retrieved from https://members.wto.org/crnattachments/2015/TBT/KWT/15_0062_00_e.pdf.

469 Ibid.

470 Ibid.

471 Ibid.

472 Ibid.

473 Ibid.

474 Ibid.

475 Ibid.

476 Ibid.

477 Ibid.

478 Ibid.

479 Ibid.

480 Ibid.

481 Law No. 21 of 2009 Approving the Pesticides Act in the Countries of the Cooperation Council for the Arab States of the Gulf. http://extwprlegs1.fao.org/docs/pdf/kuw100469.pdf.

482 Kuwait News Agency (2010, March 30). Kuwait pesticide registration law put in place. Retrieved from http://www.kuna.net.kw/ArticlePrintPage.aspx?id=2072227&language=en.

483 Ministry of Health (n.d.). Registration Certificate of Pesticides (Agriculture/Public Health) Required to be Imported and Used Inside Kuwait. Retrieved from https://www.e.gov.kw/sites/kgoenglish/Pages/Services/MOH/RegistrationOfPesticides.aspx.

484 Ibid.

485 Ibid.

486 Ibid.

487 Ibid.

488 Ibid.

489 Ibid.

490 Garlaparti, A. and Adivi, B. (n.d.). An Outline of HSE Regulations of State of Kuwait. American Society of Safety Engineers (ASSE). Retrieved from http://www.assekuwait.org/download/books/1904151047181585.pdf.

491 Ibid.

492 Ibid.

493 Anonymous (n.d.). Kuwait – 2013. International Labor Organization. Retrieved from http://www.ilo.org/dyn/legosh/en/f?p=LEGPOL:1100:6298659833438::::P1100_THEME_ID:100500.

494 Garlaparti, A. and Adivi, B. (n.d.). An Outline of HSE Regulations of State of Kuwait. American Society of Safety Engineers (ASSE). Retrieved from http://www.assekuwait.org/download/books/1904151047181585.pdf.

495 Ibid.

496 Ibid.

497 Attar, Z. (2013, September 26). Hazard Communication for the Middle East and Africa. Retrieved from https://schc.memberclicks.net/assets/meetings/fall2013/zeina_attar.ppt.pdf.

498 Alsulaili, A., AlSager, B., Albanwan, H., Almeer, A. and AlEssa, A. (2014). An Integrated Solid Waste Management System in Kuwait. 5th International Conference on Environmental Science and Technology, Vol. 69. IACSIT Press, Singapore.

499 Attar, Z. (2013, September 26). Hazard Communication for the Middle East and Africa. Retrieved from https://schc.memberclicks.net/assets/meetings/fall2013/zeina_attar.ppt.pdf.

500 Ibid.

501 Anonymous (2017, July 24). Kuwait – Labeling/Marking Requirements. Export.gov. Retrieved from https://www.export.gov/article?id=Kuwait-Labeling-Marking-Requirements.

502 Sultanate of Oman (n.d.). Retrieved from http://www.omansultanate.com/.

503 Oman (2017). CIA World Factbook. Retrieved from https://www.cia.gov/library/publications/the-world-factbook/geos/mu.html.

504 Ibid.

505 Ibid.

506 Ibid.

507 Ibid.

508 Ibid.

509 Royal Decree No. 84/2011 – Issuing the Law of the Official Gazette. Ministry of Legal Affairs.

510 Anonymous (n.d.). Regional Workshop on Chemical Hazard Communication and GHS Implementation for Arab Countries. Retrieved from http://cwm.unitar.org/publications/publications/event/ghs_arab_workshop_2006/Arab_Regional_GHS_Workshop_Report_final_rev.pdf.

511 Ibid.

512 Anonymous (n.d.). Laws and Legislation. Sustainable Oman. Retrieved from http://www.sustainableoman.com/legislation/.

513 Ibid.

514 Ibid.

515 The Law of Handling and Use of Chemicals. Official Gazette (1995, October 1). Retrieved from https://www.ilo.org/dyn/natlex/docs/ELECTRONIC/83512/92283/F456094927/OMN83512.pdf.

516 Ibid.

517 Ibid.

518 Ibid.

519 Ibid.

520 Ibid.

521 Ibid.

522 Ibid.

523 Ministerial Decision No. 248/1997. Issuing the Regulation for the Registration of Chemical Substances and the Relevant Permits. Retrieved from http://www.pdo.co.om/hseforcontractors/Environment/Documents/Oman%20Laws/Misterial%20Decision%20-%20Guidelines/Registration%20of%20Chemical%20Substances%20and%20the%20Relevant%20Permits%20.pdf.

524 Ibid.

525 Ibid.

526 Ibid.

527 Ibid.

528 Ministerial Decree No. 25/2009 "Issuing the Regulation for Organization of Handing and Use of Chemicals. Retrieved from http://www.sustainableoman.com/wp-content/uploads/2016/05/MD-25-2009e-Regulations-for-Organization-of-Handling-and-Use-of-Chemicals.pdf.

529 Ibid.

530 Ibid.

531 Ibid.

532 Ibid.

533 Royal Decree No. 64/2006 "Issuing the Pesticide Law." Food and Agriculture Organization of the United States. Retrieved from http://www.fao.org/faolex/results/details/en/c/LEX-FAOC156685/.

534 Ibid.

535 Ibid.

536 Ibid.

537 Ibid.

538 Ibid.

539 Ministerial Decree No. 41/2012 "Issuing the Legal Rule for the Pesticides Law." Retrieved from http://www.fao.org/faolex/results/details/en/c/LEX-FAOC131192/.

540 Ibid.

541 Ibid.

542 Anonymous (2012, February 15). Health and Safety in the Workplace. Oman Law Blog. Retrieved from http://omanlawblog.curtis.com/2012/02/health-and-safety-in-workplace.html.

543 Ibid.

544 Ibid.

545 Ibid.

546 Ibid.

547 Ibid.

548 Ibid.

549 Ibid.

550 Ibid.

551 Anonymous (n.d.). Laws and Legislation. Sustainable Oman. Retrieved from http://www.sustainableoman.com/legislation/.

552 Ministerial Decision No. 18-93 "Regulations for the Management of Hazardous Waste." Retrieved from https://meca.gov.om/en/files/decisions/Decisions6677798423.pdf/.

553 Ibid.

554 Ibid.

555 Ibid.

556 Ibid.

557 Ibid.

558 Ibid.

559 Ibid.

560 Ibid.

561 Ibid.

562 Szeps-Znaider, T. (2009, November 5). Chemical Control Legislation in the Middle East: Varied and Evolving. *Environmental Business Journal*, XXII(5), pp. 9–12.

563 Ministerial Decision No. 248/1997. Issuing the Regulation for the Registration of Chemical Substances and the Relevant Permits. Retrieved from http://www.pdo.co .om/hseforcontractors/Environment/Documents/Oman%20Laws/Misterial %20Decision%20-%20Guidelines/Registration%20of%20Chemical%20Substances %20and%20the%20Relevant%20Permits%20.pdf.

564 Szeps-Znaider, T. (2009, November 5). Chemical Control Legislation in the Middle East: Varied and Evolving. *Environmental Business Journal*, XXII(5), pp. 9–12.

565 Ministerial Decision No. 317/2001. Issuing the Regulations for the Packing, Packaging, and Labeling of Hazardous Chemical. Retrieved from http://www.pdo.co.om/ hseforcontractors/Environment/Documents/Oman%20Laws/Misterial%20Decision %20-%20Guidelines/ Packing,%20Packaging,%20and%20Labeling%20of%20Hazardous%20Chemicals.pdf.

566 Ibid.

567 Ibid.

568 Ibid.

569 Ibid.

570 Pakistan (2017). CIA World Factbook. Retrieved from https://www.cia.gov/library/ publications/the-world-factbook/geos/pk.html.

571 Ibid.

572 Ibid.

573 Anonymous (2006). Pakistan. Guide to Foreign and International Legal Citations. *Journal of International Law and Politics*. New York University School of Law. Retrieved from http://www.sa.i-pdf.info/h-political/47858-19-journal-international-law-and-politics-guide-foreign-and-internation.php.

574 Ibid.

575 Ibid.

576 Ibid.

577 Ibid.

578 Pakistan (2017). CIA World Factbook. Retrieved from https://www.cia.gov/library/ publications/the-world-factbook/geos/pk.html.

579 Ibid.

580 Anonymous (2006). Pakistan. Guide to Foreign and International Legal Citations. *Journal of International Law and Politics*. New York University School of Law. Retrieved from http://www.sa.i-pdf.info/h-political/47858-19-journal-international-law-and-politics-guide-foreign-and-internation.php.

581 Pakistan (2017). CIA World Factbook. Retrieved from https://www.cia.gov/library/ publications/the-world-factbook/geos/pk.html.

582 Ibid.

583 Ibid.

584 Ibid.

585 Anonymous (2006). Pakistan. Guide to Foreign and International Legal Citations. *Journal of International Law and Politics*. New York University School of Law. Retrieved from http://www.sa.i-pdf.info/h-political/47858-19-journal-international-law-and-politics-guide-foreign-and-internation.php.

586 Gillani, S.M. (n.d.). Gazette of Pakistan. Retrieved from https://www.scribd.com/presentation/205558646/Gazette-of-Pakistan.

587 Ibid.

588 National Capacity Assessment of Pakistan for Implementation of Strategic Approach to International Chemical Management (SAICM), op. cit., p. 9.

589 Pakistan Environmental Protection Act. Retrieved from http://www.environment.gov.pk/act-rules/envprotact1997.pdf.

590 Ibid.

591 Ibid.

592 Jaspal, Z.N. and Haider, N. (2014). Management of Chemicals in Pakistan: Concerns and Challenges. *South Asian Studies*, 29(2), pp. 497–517.

593 Ibid.

594 Hazardous Substances Rules (2014). Retrieved from http://sindhlaws.gov.pk/setup/Publications_SindhCode/PUB-15-000256.pdf.

595 Ibid.

596 Ibid.

597 Ibid.

598 Ibid.

599 Ibid.

600 Ibid.

601 Ibid.

602 Ibid.

603 Ibid.

604 Ibid.

605 Ibid.

606 Ibid.

607 Ibid.

608 Ibid.

609 Ibid.

610 Ibid.

611 Ibid.

612 Ibid.

613 Pakistan Environmental Protection Act (1997). Retrieved from http://www.environment.gov.pk/act-rules/envprotact1997.pdf.

614 Ibid.

615 Gillani, S.Z.A. (n.d.). Implementation of Basel Convention and Other International Chemicals/Wastes Management Instruments in Pakistan [PowerPoint slides]. Retrieved from www.archive.basel.int/ships/ktw/10.ppt.

616 Pakistan Environmental Protection Act (1997). Retrieved from http://www.environment.gov.pk/act-rules/envprotact1997.pdf.

617 Ibid.

618 Ibid.

619 Ibid.

620 Ibid.

621 Handling, Manufacture, Storage, Import of hazardous waste and hazardous substances Rules, 2016. Government of Pakistan, Ministry of Climate Change. Retrieved from www.environment.gov.pk/act-rules/20160201CORRECTEDHAZARDOUS2016.doc.

622 Ibid.

623 Ibid.

624 Ibid.

625 Ibid.

626 Ibid.

627 Ibid.

628 Ibid.

629 Ibid.

630 Ibid.

631 Ibid.

632 Ibid.

633 Ibid.

634 Ibid.

635 Ibid.

636 Ibid.

637 Ibid.

638 Ibid.

639 Ibid.

640 Ibid.

641 Filbin. A. (2015, November 30). Is It Flammable? THE New OSHA Flammable Liquids Definition. The Thrival Company. Retrieved from http://thrivalschool.com/is-it-flammable-the-new-osha-flammable-liquids-definition/.

642 Handling, Manufacture, Storage, Import of hazardous waste and hazardous substances Rules, 2016. Government of Pakistan, Ministry of Climate Change. Retrieved from www.environment.gov.pk/act-rules/20160201CORRECTEDHAZARDOUS2016.doc.

643 Guidance for Hazard Determination for Compliance with the OSHA Hazard Communication Standard. United States Department of Labor, Occupational Safety and Health Administration. Retrieved from https://www.osha.gov/dsg/hazcom/ghd053107.html.

644 Handling, Manufacture, Storage, Import of hazardous waste and hazardous substances Rules, 2016. Government of Pakistan, Ministry of Climate Change. Retrieved from www.environment.gov.pk/act-rules/20160201CORRECTEDHAZARDOUS2016.doc.

645 Ibid.

646 Anonymous (2016, June 13). Pesticides Registration and Regulation in Pakistan [PowerPoint slides]. Retrieved from http://cac-pakistan.com/Uploads/Editor/2016-6-13/Presentation%20on%20pesticides%20Registration%20&%20Regulation%20in%20Pakistan.pdf.

647 Ibid.

648 Ibid.

649 Ibid.

650 Ibid.

651 Ibid.

652 The Agricultural Pesticides Ordinance, 1971.

653 Ibid.

654 Ibid.

655 Ibid.

656 Ibid.

657 Ibid.

658 Ibid.

659 Ibid.

660 Ibid.

661 Rizvi, A. (2013). Requirements to Sell, Manufacture or Commercialize Transgenics, Insecticides, Pesticides, Herbicides and Rodenticides. LexMundi. Retrieved from http://www.lexmundi.com/document.asp?docid=917.

662 The Agricultural Pesticides Ordinance, 1971.

663 Ibid.

664 Anonymous (2016, June 13). Pesticides Registration and Regulation in Pakistan [PowerPoint slides]. Retrieved from http://cac-pakistan.com/Uploads/Editor/2016-6-13/Presentation%20on%20pesticides%20Registration%20&%20Regulation%20in%20Pakistan.pdf.

665 Ibid.

666 Ibid.

667 Ibid.

668 Ibid.

669 Ibid.

670 Ibid.

671 Ibid.

672 Ibid.

673 Ibid.

674 Anonymous (2014, January 10). Pakistan: First Information Reports (FIRs) (2010-December 2013). Immigration and Refugee Board of Canada. Retrieved from http://www.refworld.org/docid/52eba0d84.html.

675 Anonymous (2016, June 13). Pesticides Registration and Regulation in Pakistan [PowerPoint slides]. Retrieved from http://cac-pakistan.com/Uploads/Editor/2016-6-13/Presentation%20on%20pesticides%20Registration%20&%20Regulation%20in%20Pakistan.pdf.

676 Anonymous (n.d.). Occupational Safety and Health Laws in Pakistan. Retrieved from http://www.paycheck.pk/main/labour-laws/health-safety-at-work/occupational-safety-and-health/occupational-safety-and-health-laws-in-pakistan.

677 Shaw, R. (2015, December 21). An Overview of Occupational Health and Safety Laws and Regulations in Asia and the Middle East. Redlog Environmental. Retrieved from http://www.redlogenv.com/worker-safety/an-overview-of-worker-health-and-safety-laws-and-regulations-in-asia.

678 Anonymous (n.d.). Occupational Safety and Health Laws in Pakistan. Retrieved from http://www.paycheck.pk/main/labour-laws/health-safety-at-work/occupational-safety-and-health/occupational-safety-and-health-laws-in-pakistan.

679 Ibid.

680 Ibid.

681 Ibid.

682 Ibid.

683 National Capacity Assessment of Pakistan for Implementation of Strategic Approach to International Chemical Management (SAICM), op. cit., p. 9.

684 Ibid.

685 Ibid.

686 Hazardous Substances Rules, 2014. Retrieved from http://sindhlaws.gov.pk/setup/ Publications_SindhCode/PUB-15-000256.pdf.

687 Ibid.

688 Ibid.

689 Szeps-Znaider, T. (2009, November 5). Chemical Control Legislation in the Middle East: Varied and Evolving. *Environmental Business Journal*, XXII(5), pp. 9–12.

690 Ibid.

691 Pakistan Industrial Label Review (n.d.). Retrieved from http://www.nexreg.com/label-services/pakistan-industrial-label-review.

692 Ibid.

693 Saudi Arabia (2017). CIA World Factbook. Retrieved from https://www.cia.gov/library/publications/the-world-factbook/geos/sa.html.

694 Anonymous (2016, July 11). Saudi Arabia – Trade Standards. Retrieved from https://www.export.gov/article?id=Saudi-Arabia-trade-standards.

695 Saudi Arabia (2017). CIA World Factbook. Retrieved from https://www.cia.gov/library/publications/the-world-factbook/geos/sa.html.

696 Ibid.

697 Anonymous (2016, July 11). Saudi Arabia – Trade Standards. Retrieved from https://www.export.gov/article?id=Saudi-Arabia-trade-standards.

698 Ibid.

699 https://verigates.bureauveritas.com/wps/wcm/connect/68534677-0e1b-4c18-af27-e2d8ebace368/SAUDI+ARABIA++-+Import+Guide+01+-+Banned+and+restricted+products+Ed.+1.8.pdf?MOD=AJPERES.

700 Ibid.

701 Piccolo, C. (n.d.). The Hejira, or Islamic Calendar. Retrieved from https://www.hziegler.com/articles/heijira-islamic-calendar.html.

702 Anonymous (2006). Saudi Arabia. Guide to Foreign and International Legal Citations. *Journal of International Law and Politics*. New York University School of Law.

703 Saudi Arabia (2017). CIA World Factbook. Retrieved from https://www.cia.gov/library/publications/the-world-factbook/geos/sa.html.

704 Anonymous (2006). Saudi Arabia. Guide to Foreign and International Legal Citations. *Journal of International Law and Politics*. New York University School of Law.

705 Ibid.

706 Saudi Arabia (2017). CIA World Factbook. Retrieved from https://www.cia.gov/library/publications/the-world-factbook/geos/sa.html.

707 Ibid.

708 Ibid.

709 Anonymous (2006). Saudi Arabia. Guide to Foreign and International Legal Citations. *Journal of International Law and Politics*. New York University School of Law.

710 Ibid.

711 Saudi Arabia Overview (n.d.). Retrieved from https://www.google.com/url?sa=t&rct=j&q=&esrc=s&source=web&cd=1&cad=rja&uact=8&ved=0CB8QFjAA&url=http%3A%2F%2Fita.doc.gov%2Ftd%2Fstandards%2FMarkets%2FAfrica%2C%2520NearEast%2520and%2520South%2520Asia%2FSaudi%2520Arabia%2FSaudi%2520Arabia.pdf&ei=wlM2Vcz6Ee2xsATInYD4DQ&usg=AFQjCNHwPkLCJIDbdX2lNvYXZQEzR0XImw&sig2=dRsbcu7Gu_qssd06nNRVEg.

712 Ibid.

713 Anonymous (2016, July 11). Saudi Arabia – Trade Standards. Retrieved from https://www.export.gov/article?id=Saudi-Arabia-trade-standards.

714 Ibid.

715 Ibid.

716 Attar, Z. (2013, September 26). Hazard Communication for the Middle East and Africa. Retrieved from https://schc.memberclicks.net/assets/meetings/fall2013/zeina_attar.ppt.pdf.

717 Ibid.

718 Anonymous (2014, May). Saudi Arabia Environment, Health & Safety Profile and Checklist. The Isosceles Group. Retrieved from https://nimonik.com/case_studies/audit_previews/ksa_audit_preview_health_safety_environment.pdf.

719 Law of Chemicals Import and Management (2006, June 12). Royal Decree No. M/38. Retrieved from https://www.boe.gov.sa/ViewSystemDetails.aspx?lang=en&SystemID=251&VersionID=79.

720 Ibid.

721 Ibid.

722 Ibid.

723 Ibid.

724 Ibid.

725 Anonymous (2014, May). Saudi Arabia Environment, Health & Safety Profile and Checklist. The Isosceles Group. Retrieved from https://nimonik.com/case_studies/audit_previews/ksa_audit_preview_health_safety_environment.pdf.

726 Ibid.

727 Anonymous (n.d.). Regional Workshop on Chemical Hazard Communication and GHS Implementation for Arab Countries. Retrieved from http://cwm.unitar.org/publications/publications/event/ghs_arab_workshop_2006/Arab_Regional_GHS_Workshop_Report_final_rev.pdf.

728 Anonymous (2016, July 11). Saudi Arabia – Trade Standards. Retrieved from https://www.export.gov/article?id=Saudi-Arabia-trade-standards.

729 Szeps-Znaider, T. (2009, November 5). Chemical Control Legislation in the Middle East: Varied and Evolving. *Environmental Business Journal*, XXII(5), pp. 9–12.

730 Attar, Z. (2013, September 26). Hazard Communication for the Middle East and Africa. Retrieved from https://schc.memberclicks.net/assets/meetings/fall2013/zeina_attar.ppt.pdf.

731 Szeps-Znaider, T. (n.d.). Chemical Regulation in the Middle East – An Overview. Chemical Watch.

732 Ibid.

733 Anonymous (2004). Royal Commission Environmental Regulations, v.1. Kingdom of Saudi Arabia – Royal Commission for Jubail and Yanbu. Environmental Control Department. Retrieved from https://www.slideshare.net/sudeebkumar/royal-commission-jubail-environmental-regulations-2015volume-i.

734 Ibid.

735 Attar, Z. (2013, September 26). Hazard Communication for the Middle East and Africa. Retrieved from https://schc.memberclicks.net/assets/meetings/fall2013/zeina_attar.ppt.pdf.

736 Ibid.

737 Ibid.

738 Ibid.

739 Ibid.

740 Ibid.

741 Ibid.

742 Anonymous (n.d.). Environmental Laws. Hatem Abbas Ghazzawi & Co. Retrieved from http://www.saudilegal.com/saudilaw/18_law.html.

743 Ibid.

744 Environmental Law (2001, October 15). Royal Decree No. M/34. Retrieved from https://www.boe.gov.sa/ViewSystemDetails.aspx?lang=en&SystemID=117& VersionID=143.

745 Ibid.

746 Ibid.

747 Ibid.

748 Ibid.

749 Ibid.

750 Ibid.

751 Ibid.

752 Ibid.

753 Ibid.

754 The Law of Chemical Materials Importing and its Management (n.d.). Retrieved from http://www.mci.gov.sa/en/LawsRegulations/SystemsAndRegulations/chemicals/ Pages/regulation.aspx?AspxAutoDetectCookieSupport=1.

755 Law of Chemicals Import and Management (2006, June 12). Royal Decree No. M/38. Retrieved from https://www.boe.gov.sa/ViewSystemDetails.aspx?lang=en& SystemID=251&VersionID=79.

756 Ibid.

757 Ibid.

758 Ibid.

759 Ibid.

760 Ibid.

761 Ibid.

762 Ibid.

763 Ibid.

764 Ibid.

765 Ibid.

766 Ibid.

767 Ibid.

768 Ibid.

769 The Law of Chemical Materials Importing and its Management (n.d.). Retrieved from http://www.mci.gov.sa/en/LawsRegulations/SystemsAndRegulations/chemicals/ Pages/regulation.aspx?AspxAutoDetectCookieSupport=1.

770 Law of Chemicals Import and Management (2006, June 12). Royal Decree No. M/38. Retrieved from https://www.boe.gov.sa/ViewSystemDetails.aspx?lang=en& SystemID=251&VersionID=79.

771 Ibid.

772 Chakibi, S. (2013, January 26). Saudi Arabia Releases 9 New Environmental Laws. EHS Journal. Retrieved from http://ehsjournal.org/http:/ehsjournal.org/sanaa-chakibi/saudi-arabia-9-new-environmental-laws/2013/.

773 Ibid.

774 Al-Majathoub, M. (2015, December 16). Pesticide Market Access & Regulatory Management in Saudi Arabia. Retrieved from http://slideplayer.com/slide/4323061/.

775 Ibid.

776 Anonymous (2009, February 27). Officials ban a number of pesticides from market. Retrieved from http://www.arabnews.com/node/321444.

777 Al-Majathoub, M. (2015, December 16). Pesticide Market Access & Regulatory Management in Saudi Arabia. Retrieved from http://slideplayer.com/slide/4323061/.

778 https://verigates.bureauveritas.com/wps/wcm/connect/68534677-0e1b-4c18-af27-e2d8ebace368/SAUDI+ARABIA++-+Import+Guide+01+-+Banned+and+restricted+products+Ed.+1.8.pdf?MOD=AJPERES.

779 Ibid.

780 Al-Majathoub, M. (2015, December 16). Pesticide Market Access & Regulatory Management in Saudi Arabia. Retrieved from http://slideplayer.com/slide/4323061/.

781 Ibid.

782 Ibid.

783 Ibid.

784 Ibid.

785 Ibid.

786 Ibid.

787 Ibid.

788 Ibid.

789 Ibid.

790 Anonymous (n.d.). Collaborative International Pesticides Analytical Council. Retrieved from http://www.cipac.org/.

791 Anonymous (n.d.). Private Laboratories – Regulations. Royal Embassy of Saudi Arabia. Retrieved from https://saudiembassy.net/private-laboratories-regulations.

792 Ibid.

793 Ibid.

794 Ibid.

795 Ibid.

796 Ibid.

797 Khoja, S. and Aljoaid, N. (2015, November 18). Workers' Health and Safety Regulations in the Kingdom of Saudi Arabia. Clyde & Co. Retrieved from https://www.clydeco.com/insight/article/workers-health-and-safety-regulations-in-the-kingdom-of-saudi-arabia.

798 Ibid.

799 Mahayni, Z. and Mahayni, Z. (2016, May 6). Saudi Arabia: The New Implementing Regulations to the Saudi Arabian Labor Law. Mondaq. Retrieved from http://www.mondaq.com/saudiarabia/x/489208/employee+rights+labour+relations/The+New+Implementing+Regulations+To+The+Saudi+Arabian+Labor+Law.

800 Anonymous (2014, May). Saudi Arabia Environment, Health & Safety Profile and Checklist. The Isosceles Group. Retrieved from https://nimonik.com/case_studies/audit_previews/ksa_audit_preview_health_safety_environment.pdf.

801 Labor Law. Retrieved from
https://saudiembassy.net/labor-and-workmen-law#Chapter One: General Provisions.

802 Ibid.

803 Mahayni, Z. and Mahayni, Z. (2016, May 6). Saudi Arabia: The New Implementing
Regulations to the Saudi Arabian Labor Law. Mondaq. Retrieved from http://www
.mondaq.com/saudiarabia/x/489208/employee+rights+labour+relations/The+New+
Implementing+Regulations+To+The+Saudi+Arabian+Labor+Law.

804 Ibid.

805 Anonymous (n.d.). Saudi Arabia: Waste Transportation, Standard, 2012. Retrieved
from http://www.complianceandrisks.com/regulations/saudi-arabia-waste-
transportation-standard-2012-17821/.

806 Ibid.

807 Chakibi, S. (2013, January 26). Saudi Arabia Releases 9 New Environmental Laws. EHS
Journal. Retrieved from http://ehsjournal.org/http:/ehsjournal.org/sanaa-chakibi/
saudi-arabia-9-new-environmental-laws/2013/.

808 Saudi Arabia Overview (n.d.). Retrieved from https://www.google.com/url?sa=t&
rct=j&q=&esrc=s&source=web&cd=1&cad=rja&uact=8&ved=0CB8QFjAA&url=http
%3A%2F%2Fita.doc.gov%2Ftd%2Fstandards%2FMarkets%2FAfrica%2C
%2520NearEast%2520and%2520South%2520Asia%2FSaudi%2520Arabia%2FSaudi
%2520Arabia.pdf&ei=wlM2Vcz6Ee2xsATInYD4DQ&
usg=AFQjCNHwPkLCJIDbdX2lNvYXZQEzR0XImw&sig2=dRsbcu7Gu_
qssd06nNRVEg.

809 Anonymous (2006, July 11). Saudi Arabia – Trade Standards. Retrieved from https://
www.export.gov/article?id=Saudi-Arabia-trade-standards.

810 Ibid.

811 CEFIC (2011, May). Country Sheet: Global Emerging Regulations Issue
Team – Middle East. Retrieved from www.coatings.org.uk/Media/Download.aspx?
MediaId=2366.

812 Law of Chemicals Import and Management (2006, June 12). Royal Decree No. M/38.
Retrieved from https://www.boe.gov.sa/ViewSystemDetails.aspx?lang=en&
SystemID=251&VersionID=79.

813 CEFIC (2011, May). Country Sheet: Global Emerging Regulations Issue
Team – Middle East. Retrieved from www.coatings.org.uk/Media/Download.aspx?
MediaId=2366.

814 United Arab Emirates Country Brief. Australian Government Department of Foreign
Affairs and Trade. Retrieved from http://dfat.gov.au/geo/united-arab-emirates/pages/
united-arab-emirates-country-brief.aspx.

815 United Arab Emirates (2017). CIA World Factbook. Retrieved from https://www.cia
.gov/library/publications/the-world-factbook/geos/ae.html.

816 Ibid.

817 United Arab Emirates Country Brief. Australian Government Department of Foreign
Affairs and Trade. Retrieved from http://dfat.gov.au/geo/united-arab-emirates/pages/
united-arab-emirates-country-brief.aspx.

818 United Arab Emirates (2017). CIA World Factbook. Retrieved from https://www.cia
.gov/library/publications/the-world-factbook/geos/ae.html.

819 Ibid.

820 Ibid.

821 Ibid.

822 Ibid.

823 Anonymous (2016, July 18). Ministry passes new regulations on pesticides. Retrieved from https://www.thenational.ae/uae/environment/ministry-passes-new-regulations-on-pesticides-1.182318.

824 United Arab Emirates (2017). CIA World Factbook. Retrieved from https://www.cia.gov/library/publications/the-world-factbook/geos/ae.html.

825 Ibid.

826 About the Ministry. United Arab Emirates Ministry of Climate Change and Environment. Retrieved from https://www.moccae.gov.ae/en/about-ministry/about-the-ministry.aspx.

827 Ministry of Environment and Water. Government of Ras Al Khaimah. Retrieved from https://www.rak.ae/wps/portal/rak/home/edirectory/federal-entities/ministry-of-environment-water/ministry%20of%20environment%20and%20water/!ut/p/z1/xZJBc4IwEIX_Cj14ZJJAhHBEa3Vsaad1ROHCBAiaFgKmGSz_vtGbzgj21NySfS-zu-8DMdiCWNCW76jitaClvkexk2CELMvDKIDBO4b-hLxi-yVEZDYGIYhBnAnVqD2IJP1K2I4mWS0UE6rkqaSyS5gYQV0yWM4ly1QtuxEsWM4kLQ0t44qz7xGsuODfSnZGXejXlstaVLpqUJEbR6qYvENyaqbJeA4iXGDbzUhhYmZbJkYpMamdFmbhOB4uXOSmngU2fdPNZxjE_cNf-QkkHvSXKxSG0wlakDv98Mbx4ZA_0n73psBzwabl7AjWopaVDnP1x_UsIFgOtaABsWQwDXb6Z6r2JhdFDbbDQWkf_zwcYl_Tc6LlR4Ht_-OzOW1oAIlLwfzNnejMXTJ-RDqyZ3gtuIRi_eT0C87UnAU9WKyoBE1VEbvj3IyW7bGzy7aZfSyO_sMvNPVpXw!!/dz/d5/L0lHSkovd0RNQUxrQUVnQSEhLzROVkUvZW4!/.

828 Ibid.

829 Ibid.

830 Ibid.

831 Federal Law No. 24 (1999). For the Protection and Development of the Environment. Retrieved from http://extwprlegs1.fao.org/docs/pdf/uae67811E.pdf.

832 Ibid.

833 Ibid.

834 Ibid.

835 Ibid.

836 Ibid.

837 Ibid.

838 Ibid.

839 Ibid.

840 Attar, Z. (2013, September 26). Hazard Communication for the Middle East and Africa. Retrieved from https://schc.memberclicks.net/assets/meetings/fall2013/zeina_attar.ppt.pdf.

841 Anonymous (2016, March). Global Monitoring Highlights. The ENHESA Flash. Retrieved from http://www.enhesa.com/flash/global-monitoring-highlights-sept2016#middle_east.

842 Ibid.

843 Hadi, Z. (n.d.). Requirements to Sell, Manufacture or Commercialize Transgenics, Insecticides, Pesticides, Herbicides and Rodenticides. Afridi & Angell. LexMundi. Retrieved from http://www.lexmundi.com/Document.asp?DocID=927.

844 Federal Law No. (41) of the Year 1992 Concerning Pesticides. Ministry of Climate Change & Environment. Retrieved from https://www.moccae.gov.ae/assets/download/f23a98df/federal41_1992%20_e.pdf.aspx.

845 Ibid.

846 Ibid.

847 Ibid.

848 Pesticides Import Permit (active ingredient). Ministry of Climate Change & Environment. Retrieved from https://www.moccae.gov.ae/en/our-services/agricultural/permits/permit-of-pesticide-import.aspx.

849 Pesticides Registration Certificate. Ministry of Climate Change & Environment. Retrieved from https://www.moccae.gov.ae/en/our-services/environmental/certificates/registration-of-a-pesticide.aspx.

850 Release of Pesticides Consignment (active ingredient). Ministry of Climate Change & Environment. Retrieved from https://www.moccae.gov.ae/en/our-services/agricultural/permits/release-of-an-imported-pesticide-consignment.aspx.

851 Inquiring About the Required Documents for Registering a New Pesticide. Retrieved from https://www.moccae.gov.ae/en/search.aspx?type=menus|news|events|faq|circulars|albums|videos|service&query=Pesticides%20Act%20of%20Cooperation%20Council%20for%20the%20Arab%20States%20of%20the%20Gulf.

852 Customer Happiness Charter. Ministry of Climate Change & Environment. Retrieved from https://www.moccae.gov.ae/en/our-services/happiness_eq_englishaspx.aspx.

853 Customer Happiness Center. Ministry of Climate Change & Environment. Retrieved from https://www.moccae.gov.ae/en/customer-service-centers.aspx.

854 Anonymous (2016, July 18). Ministry passes new regulations on pesticides. Retrieved from https://www.thenational.ae/uae/environment/ministry-passes-new-regulations-on-pesticides-1.182318.

855 Ministerial Decision No. (849) for the Year 2010 on the Amendment of the Ministerial Decision No. (554) for the Year 2009 Concerning the Prohibited and Restricted Use Pesticides in the United Arab Emirates. Ministry of Environment and Water. Retrieved from https://www.moccae.gov.ae/assets/download/b4325cd1/Rule849_2010_e.pdf.aspx.

856 Ibid.

857 Ministry of Environment and Water Issues New Resolution on Registration and Import of Pesticides. United Arab Emirates – The Cabinet. Retrieved from https://uaecabinet.ae/en/details/news/ministry-of-environment-and-water-issues-new-resolution-on-registration-and-import-of-pesticides.

858 Wam (2015, December 14). New UAE Rules on Import of Pesticides. Emirates 24/7 News. Retrieved from http://www.emirates247.com/news/emirates/new-uae-rules-on-import-of-pesticides-2015-12-14-1.613865.

859 Ministry of Environment and Water Issues New Resolution on Registration and Import of Pesticides. United Arab Emirates – The Cabinet. Retrieved from https://uaecabinet.ae/en/details/news/ministry-of-environment-and-water-issues-new-resolution-on-registration-and-import-of-pesticides.

860 Ibid.

861 Wam (2015, December 14). New UAE Rules on Import of Pesticides. Emirates 24/7 News. Retrieved from http://www.emirates247.com/news/emirates/new-uae-rules-on-import-of-pesticides-2015-12-14-1.613865.

862 Ministry of Environment and Water Issues New Resolution on Registration and Import of Pesticides. United Arab Emirates – The Cabinet. Retrieved from https://uaecabinet.ae/en/details/news/ministry-of-environment-and-water-issues-new-resolution-on-registration-and-import-of-pesticides.

863 Wam (2015, December 14). New UAE Rules on Import of Pesticides. Emirates 24/7 News. Retrieved from http://www.emirates247.com/news/emirates/new-uae-rules-on-import-of-pesticides-2015-12-14-1.613865.

864 Ibid.

865 Ministry of Environment and Water Issues New Resolution on Registration and Import of Pesticides. United Arab Emirates – The Cabinet. Retrieved from https://uaecabinet.ae/en/details/news/ministry-of-environment-and-water-issues-new-resolution-on-registration-and-import-of-pesticides.

866 Anonymous (2016, September). Global Monitoring Highlights. The ENHESA Flash. Retrieved from http://www.enhesa.com/flash/global-monitoring-highlights-sept2016#middle_east.

867 Anonymous (2016, July 18). Ministry Passes New Regulations on Pesticides. Retrieved from https://www.thenational.ae/uae/environment/ministry-passes-new-regulations-on-pesticides-1.182318.

868 Ibid.

869 Anonymous (2016, September). Global Monitoring Highlights. The ENHESA Flash. Retrieved from http://www.enhesa.com/flash/global-monitoring-highlights-sept2016#middle_east.

870 Kenrick, V. (2012, January 31). PPE and Construction Safety in the Middle East. EHS Today. Retrieved from http://ehstoday.com/ppe/construction-safety-in-the-middle-east-0131.

871 Ibid.

872 Description of OSH Regulatory Framework. International Labor Organization. Retrieved from http://www.ilo.org/dyn/legosh/en/f?p=14100:1100:0::NO::P1100_ISO_CODE3,P1100_YEAR:ARE,2013.

873 Ibid.

874 UAE Labor Law and its Amendments. Retrieved from https://www.google.com/url?sa=t&rct=j&q=&esrc=s&source=web&cd=3&cad=rja&uact=8&ved=0ahUKEwiIl_fW4bHXAhUKNSYKHTaiCycQFggzMAI&url=https%3A%2F%2Fwww.ilo.org%2Fdyn%2Fnatlex%2Fdocs%2FELECTRONIC%2F11956%2F69376%2FF417089305%2FARE11956.pdf&usg=AOvVaw3h6tm3jZQDEd9nGPda-OsF.

875 Ibid.

876 Ibid.

877 Ibid.

878 Ibid.

879 Ibid.

880 Ibid.

881 Ibid.

882 Ibid.

883 Ibid.

884 Ibid.

885 Ibid.

886 Anonymous (2017, June 20). UAE – RoHS Regulation 2017. TűvRheinland. Retrieved from https://www.tuv.com/en/japan/about_us_jp/regulations_and_standard_updates_jp/latest_regulations_2/latest_regulations_content_325958.html.

887 Ibid.

888 Nasr, L. (2017, May 24). UAE Publishes RoHS Regulation. AssentBLOG. Retrieved from https://blog.assentcompliance.com/index.php/uae-publishes-rohs-regulation/.

889 Ibid.

890 Attar, Z. (2013, September 26). Hazard Communication for the Middle East and Africa. Retrieved from https://schc.memberclicks.net/assets/meetings/fall2013/zeina_attar.ppt.pdf.

891 Anonymous (2016, November). Import Procedures Guide UAE. Saudi Export Development Authority. Retrieved from https://www.saudiexports.sa/ar/Export-Information/Documents/UAE%20Guide%20-%20En%2020161103.pdf.

892 Attar, Z. (2013, September 26). Hazard Communication for the Middle East and Africa. Retrieved from https://schc.memberclicks.net/assets/meetings/fall2013/zeina_attar.ppt.pdf.

893 Standard Operating Procedure for licensing [sic] of Hazardous Waste Service Providers in the Emirate of Abu Dhabi. Tadweer Waste Treatment LLC. Retrieved from http://www.tadweer.ae/en/ELibrary/Technical%20Guidelines/Licensing%20of%20Waste%20Electrical%20and%20Electronic%20Equipment%20(WEEE)%20Service%20Providers.pdf.

894 Anonymous (2014, April). Technical Guidance Document for Storage of Hazardous Materials. Environment Agency – Abu Dhabi. Retrieved from https://www.ead.ae/arabic/Documents/Business%20and%20Industry/Hazardous%20Materials/EAD-EQ-PCE-TG-16%20Storage%20of%20Hazardous%20Materials.pdf.

895 Ibid.

896 Webb, C. (2010, July 15). United Arab Emirates: Environmental Laws in Abu Dhabi. Al Tamimi & Company. Retrieved from http://www.mondaq.com/x/105298/Environmental+Law/Environmental+Laws+in+Abu+Dhabi.

897 Ibid.

898 Ibid.

899 Ibid.

900 Anonymous (n.d.). EHS: Introduction. Retrieved from http://www.adm.gov.ae/en/menu/index.aspx?TWVudUlEPTExNiZtbnU9UHJp.

901 Anonymous (n.d.). Abu Dhabi updates regulatory requirements for hazardous chemicals. Retrieved from https://chemicalwatch.com/11029/abu-dhabi-updates-regulatory-requirements-for-hazardous-chemicals.

902 Anonymous (n.d.). About Center. Retrieved from https://www.oshad.ae/en/Pages/Aboutcenter.aspx.

903 Anonymous (n.d.). Sectors. Retrieved from https://www.oshad.ae/en/sectors/Pages/sectors.aspx.

904 Ibid.

905 Environment Agency – Abu Dhabi (2005). Law No. 21 For Waste Management in the Emirate of Abu Dhabi. Retrieved from http://www.ilo.org/dyn/natlex/docs/ELECTRONIC/83638/92487/F1519557539/ARE83638.pdf.

906 Ibid.

907 Ibid.

908 Ibid.

909 Ibid.

910 Ibid.

911 Ibid.

912 Ibid.

913 Abed, G.T. and Davoodi, H.R. (2003). International Monetary Fund. Challenges of Growth and Globalization in the Middle East and North Africa. Retrieved from https://www.imf.org/external/pubs/ft/med/2003/eng/abed.htm.

914 Ibid.

915 Ibid.

Index

Chemical Regulation in the Middle East, First Edition. Michael S. Wenk.
© 2018 John Wiley & Sons Ltd. Published 2018 by John Wiley & Sons Ltd.